AF360919

Search Engine Optimization

Search Engine Optimization

Editors

Andreas Veglis
Dimitrios Giomelakis

MDPI • Basel • Beijing • Wuhan • Barcelona • Belgrade • Manchester • Tokyo • Cluj • Tianjin

Editors
Andreas Veglis
Aristotle University of Thessaloniki
Greece

Dimitrios Giomelakis
Aristotle University of Thessaloniki
Greece

Editorial Office
MDPI
St. Alban-Anlage 66
4052 Basel, Switzerland

This is a reprint of articles from the Special Issue published online in the open access journal *Future Internet* (ISSN 1999-5903) (available at: https://www.mdpi.com/journal/futureinternet/special_issues/SEO).

For citation purposes, cite each article independently as indicated on the article page online and as indicated below:

LastName, A.A.; LastName, B.B.; LastName, C.C. Article Title. *Journal Name* **Year**, *Article Number*, Page Range.

ISBN 978-3-03936-818-1 (Pbk)
ISBN 978-3-03936-819-8 (PDF)

Contents

About the Editors

Andreas Veglis is a professor, and is head of the Media Informatics Lab at the School of Journalism & Mass Communication at the Aristotle University of Thessaloniki. He completed his BSc in Physics, MSc in Electronics and Communications, and PhD in Computer Science at Aristotle University. His research interests include information technology in journalism, new media, course support environments, data journalism, open data and distance learning. He has published in books and international journals like *Journal of Applied Journalism & Media Studies*, *Journal of Greek Media & Culture*, *Digital Journalism*, *Studies in Media and Communication*, *Computers in Human Behavior*, *International Journal of Monitoring and Surveillance Technologies Research*, *International Journal of Advanced Computer Science and Information Technology*, *International Journal of Computers and Communications*, *International Journal of E-Politics*, *Journalism & Mass Communication Educator*, *International Communication Gazette*, *Publishing Research Quarterly*, *International Journal of Electronic Governance*, *Journalism*, *Austral Comunicación*, *Journalism Practice*, *Future Journalism*, and *Social Sciences*. He is the author or co-author of 51 books and book chapters, he has published 87 papers in scientific journals, and he has presented 134 papers at international and national conferences. Professor Veglis has been involved in 40 national and international research projects.

Dimitrios Giomelakis (Post-doc Researcher) was born in Thessaloniki, Greece, on July 30, 1985. He completed his PhD in online journalism at the Aristotle University of Thessaloniki. His thesis examined the use and impact of different technological tools and internet services on news media websites and journalistic content in general. Currently, he is a postdoctoral researcher in the Media Informatics Lab at the School of Journalism & Mass Communications, Aristotle University of Thessaloniki, Greece. He is a graduate of the School of Journalism & Mass Communications. In 2010, he completed his master's degree in Information and Communication Technologies (ICTs) for audio-visual production and education at the Polytechnic School in Aristotle University, Thessaloniki. Among his research interests are news and journalism studies, online journalism, media technology, digital media use, SEO, web analytics, web metrics, Web 2.0 and social media. His work has been published in several scientific journals, edited books and conference proceedings.

 *future internet*

Editorial

Search Engine Optimization

Andreas Veglis [1,*] and **Dimitrios Giomelakis** [1,2]

[1] Media Informatics Lab, School of Journalism & Mass Communications, Aristotle University of Thessaloniki, 541 24 Thessaloniki, Greece; dgiomela@jour.auth.gr

[2] Informatics and Telematics Institute (ITI) of the Centre of Research & Technology Hellas (CERTH), 570 01 Thessaloniki, Greece

* Correspondence: veglis@jour.auth.gr

Received: 25 December 2019; Accepted: 31 December 2019; Published: 31 December 2019

1. Introduction

The introduction of the World Wide Web (WWW), 25 years ago, has considerably altered the manner in which people obtain information. Soon after the introduction of the WWW, it was evident that traditional browsing was totally insufficient for internet users to locate the information that interests them. This need was covered with the development of search engines. Today, search engines play one of the most important roles in disseminating content. Search engine optimization (SEO) is a collection of strategies that improves a website's presence and visibility on a search engine's results page (SERP). In other words, the higher and more frequently a site appears in search results, the more visitors it will receive through the use of search engines. The importance of SEO can be understood by the fact that many web sites today receive the majority of their web traffic through a search engine's organic results.

The methods that SEO includes can be divided into four major categories: keyword research/selection, search engine indexing, on-page optimization, and off-page optimization. On-page optimization includes the management of all factors associated directly with someone's website (e.g., keywords, appropriate content, and internal link structure), while off-page optimization includes all the efforts made away from the website such as link building or social signal strategy. Undeniably, the world of search engine optimization has changed and evolved drastically over the years with a shift away from traditional ranking factors towards deeper analysis, and factors such as quality, multi-form content, and social signals. However, even though SEO has changed a lot, it remains an important part of any digital marketing strategy.

This special issue is soliciting theoretical and case studies contributions, discussing and treating challenges, state-of-the-art technology, and solutions on search engine optimization, including, but not limited to, the following themes related to SEO: Theory of SEO, different types of SEO, SEO criteria evaluation, search engines' algorithms, social media and SEO, SEO applications in various industries, SEO in media web sites, etc. Through invited and open call submissions, a total of five excellent articles have been accepted, following a rigorous review process that required a minimum of three reviews and at least one revision round for each paper.

2. Contributions

The first paper, written by Christos Ziakis, Maro Vlachopoulou, Theodosios Kyrkoudis, and Makrina Karagkiozidou [1], identifies the main factors that affect the ranking of a website in the search engines' results in order to provide enterprises and freelancers with a guide on the best techniques to maximize a website's position in search results. The paper consists of two parts. The first part performs a literature review through a collection and analysis of academic papers and the second part consists of research that was conducted manually using different phrases as case studies.

Minos-Athanasios Karyotakis, Evangelos Lamprou, Matina Kiourexidou, and Nikos Antonopoulos are the authors of the second paper [2] that aims to expand the current literature about the SEO practices by focusing on examining, via the walkthrough method, the ways that news companies allow the users to comment on their online news articles. The study investigates an extensive sample of Greek, Cypriot, and international news websites.

The third paper written by Cristòfol Rovira, Lluís Codina, Frederic Guerrero-Solé, and Carlos Lopezosa [3] investigates academic SEO by analyzing and comparing the relevance ranking algorithms employed by various academic platforms in order to identify the importance of citations received in their algorithms. The authors analyze two search engines and two bibliographic databases: Google Scholar and Microsoft Academic, on the one hand, and Web of Science and Scopus, on the other.

Andreas Giannakoulopoulos, Nikos Konstantinou, Dimitris Koutsompolis, Minas Pergantis, and Iraklis Varlamis have contributed the fourth paper of the special issue [4]. The purpose of this paper is to study the extent to which a university's academic excellence is related to the quality of its web presence. The study deals with the website quality and search engine optimization (SEO) performance of the most well-known university websites, using the Academic Ranking of World Universities (ARWU) Shanghai list as a base of reference.

Finally, the fifth paper was written by Dimitrios Giomelakis, Christina Karypidou, and Andreas Veglis [5] includes an exploratory study on the use of search engine optimization (SEO) in news websites. Through a series of semi-structured interviews with SEO and media professionals at four Greek media organizations, the study examines the familiarity of these news publishers with SEO practices, including common trends and practices inside their own newsrooms, and the perceived impact of SEO on journalism and news content.

Conflicts of Interest: The authors declare no conflict of interest.

References

1. Ziakis, C.; Vlachopoulou, M.; Kyrkoudis, T.; Karagkiozidou, M. Important Factors for Improving Google Search Rank. *Future Internet* **2019**, *11*, 32. [CrossRef]
2. Karyotakis, M.-A.; Lamprou, E.; Kiourexidou, M.; Antonopoulos, N. SEO Practices: A Study about the Way News Websites Allow the Users to Comment on Their News Articles. *Future Internet* **2019**, *11*, 188. [CrossRef]
3. Rovira, C.; Codina, L.; Guerrero-Solé, F.; Lopezosa, C. Ranking by Relevance and Citation Counts, a Comparative Study: Google Scholar, Microsoft Academic, WoS and Scopus. *Future Internet* **2019**, *11*, 202. [CrossRef]
4. Giannakoulopoulos, A.; Konstantinou, N.; Koutsompolis, D.; Pergantis, M.; Varlamis, I. Academic Excellence, Website Quality, SEO Performance: Is there a Correlation? *Future Internet* **2019**, *11*, 242. [CrossRef]
5. Giomelakis, D.; Karypidou, C.; Veglis, A. SEO inside Newsrooms: Reports from the Field. *Future Internet* **2019**, *11*, 261. [CrossRef]

 *future internet*

Article

Important Factors for Improving Google Search Rank

Christos Ziakis *, **Maro Vlachopoulou, Theodosios Kyrkoudis and Makrina Karagkiozidou**

ISEB lab, Dep. of Applied Informatics, University of Macedonia, 156 Egnatia St., 54006 Thessaloniki, Greece; mavla@uom.edu.gr (M.V.); it1421@uom.edu.gr (T.K.); it1414@uom.edu.gr (M.K.)
* Correspondence: ziakis@uom.edu.gr; Tel.: +30-693-746-2999

Received: 11 December 2018; Accepted: 18 January 2019; Published: 30 January 2019

Abstract: The World Wide Web has become an essential modern tool for people's daily routine. The fact that it is a convenient means for communication and information search has made it extremely popular. This fact led companies to start using online advertising by creating corporate websites. With the rapid increase in the number of websites, search engines had to come up with a solution of algorithms and programs to qualify the results of a search and provide the users with relevant content to their search. On the other side, developers, in pursuit of the highest rankings in the search engine result pages (SERPs), began to study and observe how search engines work and which factors contribute to higher rankings. The knowledge that has been extracted constituted the base for the creation of the profession of Search Engine Optimization (SEO). This paper consists of two parts. The first part aims to perform a literature review of the factors that affect the ranking of websites in the SERPs and to highlight the top factors that contribute to better ranking. To achieve this goal, a collection and analysis of academic papers was conducted. According to our research, 24 website characteristics came up as factors affecting any website's ranking, with the most references mentioning quality and quantity of backlinks, social media support, keyword in title tag, website structure, website size, loading time, domain age, and keyword density. The second part consists of our research which was conducted manually using the phrases "hotel Athens", "email marketing", and "casual shoes". For each one of these keywords, the first 15 Google results were examined considering the factors found in the literature review. For the measurement of the significance of each factor, the Spearman correlation was calculated and every factor was compared with the ranking of the results individually. The findings of the research showed us that the top factors that contribute to higher rankings are the existence of website SSL certificate as well as keyword in URL, the quantity of backlinks pointing to a website, the text length, and the domain age, which is not perfectly aligned with what the literature review showed us.

Keywords: SEO; search engine optimization; website ranking; ranking factors

---

1. Introduction

The emergence of the Internet, and its rapid expansion worldwide, resulted in the storage and sharing of a massive amount of data, and this data was available to every user with an internet connection. As a result, billions of websites were created, which made it hard for the average user to extract useful information from the web efficiently for a specific search. The need for an easier, more efficient way to search for information led to the development of search engines. Gradually, search engines began to assess the relevance of every website on their indexes compared to the queries provided to them by the users. They took into consideration several website characteristics and metrics and calculated the value of each website using complex algorithms. The enormous number of websites being indexed from search engines, along with the increasing competition for the first search results, led to studying and implementing various techniques in order for websites to appear more valuable in search engines. These techniques make up what is called today "Search Engine

Optimization" (SEO) and they are divided into either black hat or white hat depending on whether they violate the search engines' terms of service. It is worth mentioning that a website nowadays has to be indexed into the first page of SERPs in order to receive a sufficient amount of organic visitors. Ideally one of the first three results has to be acquired for the targeted keywords as the click through rates are 30% (1st position), 16% (2nd position), and 10% (3rd position), while the click through rate for positions past the first page of SERPs is limited to under 2% [1]. Search engines nowadays change their algorithms regularly by adding and removing factors affecting the ranking of websites: SEO is a dynamic procedure.

In this paper, a literature review, as well as a research review was conducted in order to pinpoint the main factors that affect the ranking of a website in the search engines' results. The knowledge from this paper aims to provide enterprises and freelancers with a guide on the best techniques to maximize a website's position in search results.

2. Methodology

In order to accomplish the study, we researched several scientific databases to find articles about SEO. The keyword used in our research is "Search Engine Optimization".

To choose the most relevant articles, we searched academic papers using the PRISMA methodology [2] as shown on Figure 1 on the digital libraries of (i) Science Direct, (ii) Springer, and (iii) IEEE Xplore during February and March of 2018. The first search resulted in almost 1010 citations. By applying the PRISMA methodology we retrieved 125 articles after title and abstract screening and used 23 articles as the base for the literature review.

The fact that the papers did not include some crucial factors that affect the ranking of a website lead us to search around the web, using the same keywords, for authorized websites including the factors that were not included in academic papers. Information only from trustworthy websites was extracted only after the opinions of the authors were carefully examined.

Concerning the research, we implemented an 8-step process to get the requested findings

1. We searched for primary research to highlight the top SEO factors.
2. We determined the factors that we will use in our research and the tools that will be used to extract data for every factor.
3. We determined the number of the samples, the size of every sample and the keywords that are going to be used for every sample.
4. A Google search was conducted using the keywords of every sample and the data was saved in EXCEL spreadsheets.
5. We normalized the data using a common numeric scale.
6. We conducted a statistical analysis using the Spearman correlation coefficient.
7. The average of the samples' Spearman correlation coefficients for every factor was calculated.
8. We represented the SEO factors graphically according to the significance of every factor.

More specifically, the research was conducted using three samples from which the first 15 search results were extracted. The key words that we used were "Hotel Athens", "email marketing", and "casual shoes". The first search was conducted on April 1, 2018 at 11:45 pm, the second on April 12, 2018 at 2:40 pm, and the third on April 13, 2018 at 4:00 pm. We chose Google as the search engine of the study as it is the most popular search engine with an almost 90% market share [3].

Concerning the factors that we took in consideration for the study, instead of using domain or page authority metrics we used the number of backlinks as a metric of the websites' value following the methods of former SEO studies [4–6].

The normalization of the data was made using as a base a numeric scale from 0 to 9 in order to compare the samples directly. We replaced the missing values with the average of the rest normalized data and we consider that it does not affect the outcome of the research.

We analyzed the data using the Spearman correlation coefficient. We used this method of statistical analysis because we are not sure that there is linear correlation between every SEO factor and the ranking of each website in the search results, [7] trying to prove the existence of monotony between them. However, we have to emphasize that correlation does not mean always causation, so we can make assumptions safely only for the factors that are highly correlated ($|r| > 0.5$) to the ranking of the websites. The values of the correlations are calculated with the web tool vassarstats.net [8] and for the measurement of the correlation degree we used the following scale.

$$R \leq -0.5 \text{ strong correlation}$$

$$-0.5 < r \leq -0.3 \text{ medium correlation}$$

$$-0.3 \leq r < -0.1 \text{ weak correlation}$$

We use the negative scale to assess whether a factor correlates to high ranking in the search results as higher score means numerically lower ranking position (from 15th to 1st position).

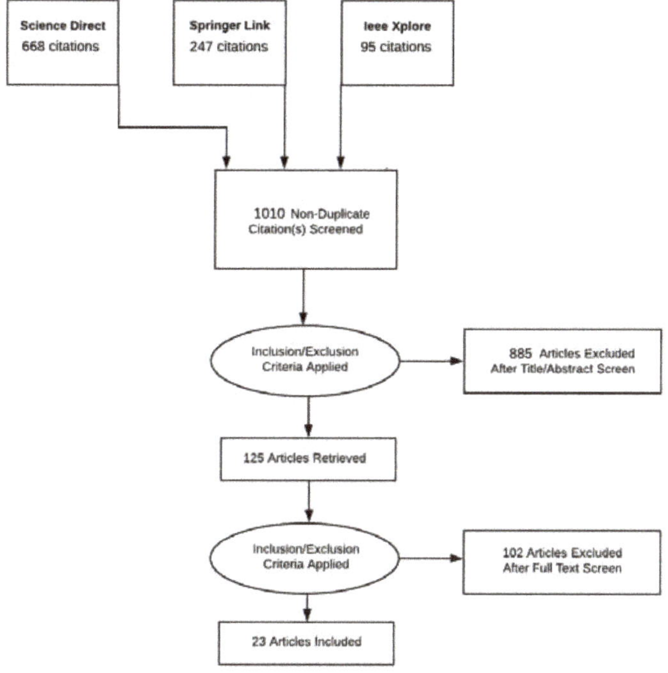

Figure 1. Application of PRISMA methodology.

3. Literature Review

We conducted a literature review to address the main factors that affect the ranking of a website as mentioned in former papers. Yalçın et al. (2010) [9], beyond the report of some factors that affect SEO, illustrated the mechanics behind the operation of the search engines. They concluded that SEO is a dynamic process which must be monitored frequently, tracing positive or negative changes for the improvement of a website. Wilson et al. (2006) [10] discussed particular SEO techniques that should be implemented during the creation of a website that aim to make the website visible on search engines. In their opinion, keywords must be related to the content of the website so it can be indexed higher in the search results for the targeted keywords. Finally, it is stressed that SEO is a competitive sector and changes are made often. Similar topics are analyzed by Zilincan (2015) [11] and Cui et al. (2011) [12]. Rehman et al. (2013) [13] evaluated the existing SEO techniques by analyzing and comparing other researches while mentioning main points that could be optimized. Using these data, they recommend some theoretical methods for SEO. This is also the main topic of the research of Zhang et al. (2011) [14] who conducted a comparative research on the SEO factors using 116 websites. Killoran (2013) [15] studies the factors that affect the ranking of the SERPs and how web developers and marketers can take advantage of them. He mentions that SERPs are shaped by three categories of participants: the search engine companies and programmers, the SEO experts, and the search engine users. It is claimed that not only the target group of the website but also the competition has to be considered while choosing keywords. Finally, he recommends that correct keyword placement and link-building through interaction with other content creators are crucial for optimal results. The same pattern has been followed by Thekral et al. (2016) [16] who focused on the theoretical documentation of the above aspects. Kakkar et al. (2015) [17] compared and explained various algorithms of Google and focuses on SEO strategies and how they lead to better rankings, concluding that SEO is a long term and dynamic procedure. Gudivada et al. (2015) [18] analyzed the mechanics behind search engines and the techniques that are used to rank websites as well as how much it impacts the traffic of a website. They conducted a study which shows that 70% of users prefer the organic results. More specific, most of them (60%) select one of the first three organic results. That proves the impact of SEO on website promotion. Krrabaj et al. (2017) [19] researched the on-site as well as the off-site SEO factors. According to the results, quality content and inbound and outbound links affect mostly the ranking of a website in the search results. Accordingly, Evans (2007) [20] made a research on the most popular SEO techniques that affect the ranking of websites in the SERPs. The sample of the research included 50 SEO optimized and 50 nonoptimized websites and the research revealed the most effective SEO techniques as well as the importance of backlinks for a successful SEO campaign. Chen et al. (2011) [21] aim to combine the SEO knowledge with the tools that the world wide web offers in order to show how to increase the traffic of a website and provide a better relationship with the website visitor. Patil Swati et al. (2013) [22] studied the ranking algorithms of the search engines and presented the main factors that affect the SEO of a website. They categorized them in "white hat" and "black hat" too, while pinpointing the differences between them. Hui et al. (2012) [23] referred analytically to the mechanics behind the operation of search engines as well as the factors that affect the ranking of websites and the method they used to select the search engine for their study. Gregurec et al. (2012) [24] briefly presented the SEO factors that have been mentioned in previous papers and they analyze some Croatian websites on the topics of computer science and engineering. They concluded that independently from the country that a website serves the same principles of SEO are applied. Chandra et al. (2014) [25] showed 32 efficient and effective methods of detecting spam in the URL, the content and the links of a website. To accomplish this, they created a classifier that is based on neural networks and requires minimal processing power. Thakur et al. (2011) [26] aim to simplify the steps for the optimization of websites to make the process easier for marketing experts. They present several SEO techniques mentioning the most flexible and effective of them. Their research constitutes a guideline for marketers in order to adopt the best techniques. Ergi et al. (2014) [27] found out in their research that the loading speed, the low bounce rate, and the total traffic of a website have a positive impact on the ranking of a website. Kumar et al. (2011) [28] study the characteristics of the algorithms that the search engines use and they suggest an optimization strategy for websites. The application of the strategy was proved effective and gave information for even better ways of website optimization. Dean (2016) [29] made a primal SEO research with the use of a crawler which crawled 1 million websites, and information about them was gathered. The main purpose of the research was to compare the ranking of the websites in the SERPs and the SEO techniques used on them. He concluded that backlinks affect mostly the position of a website followed by relevant content, the existence of at least one image as well as the existence of SSL certificate. However, they found out that techniques such as the existence of the preferred keyword in the title tag do not affect the SEO of a website. Fishkin (2017) [30], in his article, included the main factors that will affect a website's position in 2018. He provides marketers a checklist with all the data that they should take into consideration according SEO. He suggests that marketers should focus on providing value through their content while optimizing the website's structure in order to provide better user experience and content traceability. Finally, Palos-Sanchez et al. (2018) [31] focused on analyzing user behavior when using Internet applications. They found out that technology companies should stop recommending undifferentiated strategies and to adopt processes to emphasize on better user experience in a website.

4. Findings

4.1. Literature Review

Through the literature review we found that the factors as shown on Table 1 impact the ranking of websites in search engine indexes. On Table 2 the matching of previous researches to indexes is presented.

Table 1. Previous research on search engine optimization (SEO) factors.

Page Size and Website Loading Time	[7,11,21,24,26,30]
Keyword in Title tag	[8,10,11,19,20,23,25,27,30]
Keyword in H1/H2/H3 Tag	[8,19,20,30]
Keyword Density in Text	[7,8,11,19,20,25,30]
Keyword in URL	[8,19,20,30]
Keyword in Meta Description Tag	[8,15,19,20,25,30]
Alt Text	[10,13,15,23]
Unique - High Quality Content	[10,13,14,20,30]
Title Length/ Description Length	[7,9,11,19,22]
URL Length	[8,9,13,19,21,22,26,30]
Text Length	[22,26]
Text to Code Ratio	[22]
Internal Linking	[17,20,25]
Quality and Quantity of Backlinks	[7–9,12,13,15–17,20,25,26,30]
Website Structure	[7,9,14,16,19,25]
Social Media Support	[7,9,12–16,18,23,30]
Custom 404 page	[11,15,21]
SSL Certificate	[22,26,30]
Sitemap xml file	[7,8,11,15,21]
Domain age	[7–9,11,17,25]
Responsive layout	[14,15,20]
W3C Validity	[8]
Bounce Rate	[24–26]
Time on Site	[24]

The most mentioned factors in the above papers include the quality and quantity of backlinks, the social media support, the keyword in title tag, the URL length, and the website structure.

Page size and website loading time: A very important on-page factor for the optimization of a website is the loading time. Search engines include this factor in their algorithms too. The higher the loading time, the lower the ranking of a website in the search results. Although special effects and graphic elements upgrade the image of a website, the excessive use of them may increase the loading time of the website. Loading speed is affected not only by graphic elements but the HTML file and all the elements regardless of their file type contribute to the size of a website and affect the loading speed. Most search engines will not fully index pages that are greater than a certain sizeKeyword in title tag: The title tag refers to the title of the file. Every HTML/XHTML file contains a title element. The main use of the title tag is to determine the title in the toolbar of the browser, it shows the title of the website in the search results and the name in the 'favorites' tab. Concerning this factor while structuring a website it is easier for the search engines to trace and rank it in their search results.

Keyword in h1/h2/h3 tag: Equally important on page intervention is the use of keywords in h1, h2, and h3 tags. This element is the second most important on page factor for higher ranking in the search results. The websites with optimized and targeted h1, h2, and h3 tags are ranked higher by the search engines. This happens as the crawler searches for information regarding the content structure of every website and the best way to extract this piece of information is to search on these tags. Skipping this part of optimization is a big obstacle on the way to higher ranking.

Keyword density in text: The keyword density is the number of times the targeted keyword appears in the text of a website compared to the rest text. This on-page factor and the frequency of times the keyword appears in title tag are considered some of the most important factors when optimizing a website. The ideal keyword density range is 2 to 8% of the text of the page but this range is not universally true and is affected by other factors. We have to pinpoint that the keyword density should not have negative impact on the readability of the text because it is perceived as negative element of a website by the search engines

Keyword in URL: It is easier for crawlers to trace a website if its URL contains the targeted keyword. In general, search engines tend to rank higher websites with .edu and .gov domains as these domains are used by state and educational websites. However, optimization for all types of websites can be achieved if the length of the URL is kept short.

Keyword in meta description tag: The meta description tag is a summary of the content of a webpage. This tag contains the text that appears in the search results of the search engines just below the link. The meta description tag is a guide on what keywords should the website be indexed for by the search engines

Alt text: To be ensured that the most elements of a website are indexed correctly, they have to be in HTML format. However, content such as images and videos need more information to be indexed by the search engines. Webmasters must use an alt tag for images and a transcript for videos to provide enough details about those types of content.

Unique high-quality content: Search engines prefer websites with unique, authentic, and quality content. If the content of a website is plagiarized from another website it will not be ranked high by the search engines. In addition, a frequently updated content has better chance to get ranked higher in search engines for related keywords.

Title length & Description length: The title of the website must reflect the topic of the website without unnecessary information. With the use of a short and comprehensive title, website visitors can understand the main topic of the website. Google suggests no more than 70 characters for title tag. The meta tag description offers some pieces of information in the search engines about the website. A poor or no description in this tag suggests a low-quality website. On the other side, larger than standard text (up to 155 characters recommended by Google) may be considered spam as the targeted keywords may be used excessively.

URL length: The URL represents the address of the site in the World Wide Web. As mentioned above, the targeted keywords should be included in the URL, so it can be traced easily from the search engines. To become even more SEO-friendly it should be short and understandable.

Text length: Websites with longer text tend to rank higher in the search results compared to the ones that have less or no text at all. This proves that search engines prefer content rich websites. This assumption may correlate with the fact that in richer content the targeted keywords appear more frequently. Besides, users prefer longer text as it is more informative. Text to code ratio: The text-to-code ratio is a metric that represents the ratio between the text in the front-end part of the site to the back-end code. The ideal ratio range is 25 to 70%. This ratio refers in the visual text compared with all the HTML elements including the image tags and other visual elements. Although this factor is not directly correlated to the ranking of a website, there are many factors that are based on this factor, so it is crucial to be considered for a more effective SEO strategy.

Internal linking: A very important SEO factor is the linking of the individual pages of a website. The goal of internal linking is not only to make the visit of the user more enjoyable but also to facilitate the tracing and the indexing of the individual components of the website. As a result, a higher ranking of the website is achieved, given the fact that the value that is passed by the external links is shared uniformly in all subpages of the website. Finally, it should be mentioned that internal links must be checked regularly to trace any broken links.

Quality and quantity of backlinks: The quality and quantity of backlinks is one of the most important factors for optimal ranking. Backlinks are references from other websites pointing to the targeted website. Google, in order to determine the importance of each website for the user, invented an algorithm that calculates the value of the website based on the references from other websites pointing to the given site. The algorithm takes into consideration the number as well as the quality of the incoming links. This algorithm is called "Pagerank" and uses a 0 to 10 scale to determine a relative score of that page's importance and authority. The ranking of a website heavily relies on the Pagerank algorithm, which is the only factor that has remained unchanged despite the updates made in the ranking patterns. Considering the importance of this factor, building quality backlinks that are referenced from a high number of websites is extremely important to achieve high ranking. Depending on this, webmasters started implementing black hat SEO techniques such as purchasing links, arranging for links provided by link farms, and link exchange with other websites to quickly build inbound links. As soon as these techniques became known to search engine programmers, websites using these link-building strategies were banned and the algorithms were changed in order to value quality over quantity. Therefore, when building backlinks, the authority of the linking sources and the topic relevance of the linked sites must be considered.

Website structure: Having a clear navigation system is a standard requirement for a useful website. Some websites use frames and navigation buttons made in java or other programming languages which are not traceable by search engines. The solution to this problem is the creation of a complementary navigation bar using normal HTML links to ensure that every page of the website is traced by search engines. This type of navigation bar helps crawlers extract more information about the structure of the website than graphic elements. The structure of the website has to be clear as mentioned above in order to minimize the directory depth. Ideally the structure consists of less than four layers. This type of structure is achieved by embedding the most important pages, links, and titles from the second and third layer into the home page.

Social media support: It is widely known that most people use social media daily and it is the only source of information for some of them. Connecting a website to social media accounts ensures that the website gains more traffic as well as quality backlinks which increase its authority. Based on these facts, search engine algorithms begun to depend increasingly on bookmarks, social signals, and the impact of content creators on social media to assess the value of a website.

Custom 404 page: Web Servers return a 404 webpage when the requested webpage cannot be found. Customizing the 404 page helps keep users on the site and can even enhance their search experience. The customization includes pointers to home page or other pages of the site and even references to other sites with relevant to the user's search content.

SSL certificate: The acronym SSL refers to the term secure socket layer. It is a protocol that establishes an encoded link between the server and the browser that allows sensitive information to be transmitted securely. The SSL certificate ensures that a website is reliable and that the user's data are kept safe. In addition, it prevents spamming. The websites that have this certificate installed use the protocol https instead of http. Such websites tend to rank higher in the search results as they are safer than similar sites without the certification.

Sitemap xml file: An XML file that is created by the web developer of the website and is submitted for use by search engines. The creation and the uploading of a sitemap along with the website help the crawler to find all the subpages of a website. It also notifies the search engines for any changes made to the website, the degree of significance of every page, the frequency of the website updates, and other information. In sum, it contains information that enhances the effectiveness and the relevance of the indexed website's content but also makes it easier for search engines to index the website. This procedure is crucial for the optimization of a website as the number of a website's pages is highly correlated to the website's ranking.

Domain age: The age of a domain name is quite an important factor affecting the ranking of a website. According to Google's algorithms, the older the domain the more reliable the website that the domain points to. It is no surprise that websites with newer domains tend to rank lower that those with older ones.

Responsive layout: According to a 2014 study on the subject "Mobile path to Purchase" by Telemetrics and xAd [15], 50% of the respondents use a mobile device to start their search and two out of three buyers purchase goods and services through their research. Even though there is clear evidence that the mobile devices contribute to e-commerce success, many websites are not responsive to mobile devices. A responsive website not only enhances the experience of the user but also contributes to higher rankings. When making a website responsive, services that mobile devices introduced such as voice search must be considered. It is important to optimize the keywords in order to allow search using whole sentences instead of keywords.

W3C validation: W3C or the World Wide Web consortium is a worldwide web commission. This commission sets some standards regarding syntax errors in the source code of the website. Every website that meets the standards that the consortium sets ranks higher in the search engines. Websites which are validated can add a small icon that indicates that they are w3c validated.

Bounce rate: The metric that shows the percentage of visitors that leave the website just after they view the first page of the website without exploring other pages. Low bounce rate indicates a website with high quality content which is also relevant to the user's search. That is why search engines recently included that metric as a factor that affects the ranking of a website in the search results.

Time on site: The total duration of a user's visit in a website. It is obvious that the more time a user spends on a website the more valuable the content of the website is for the search engines as it is more likely for a visitor to spend more time on a page with valuable for them content. Considering the usefulness of this metric, it is becoming an increasingly more important indicator for the ranking of a website.

4.2. Our Research

After the analysis of the data for every website of the three samples that we gathered, we calculated the Spearman correlation between ranking of each website and their scores based in our scales, and then we calculated the averages for each factor and presented them concentrated using a table.

Table 2 presents all the correlations as they emerged from the samples. The columns present the samples and the rows present factors. The last row presents the average of the correlations for every factor. The averages show whether a factor is universally valid. The results are presented on Figure 2 in descending order.

We should mention that for the factors with low correlation and high score we cannot be sure whether they contribute to the ranking of websites. It is possible that the constant high score is correlated to higher ranking but we cannot confirm that as the sample of search results for every keyword is restricted to 15 webpages.

Table 2. Averages for each factor.

Seo Factors	Hotel Athens	Email Marketing	Casual Shoes	Average
Keyword in Title Tag	0.0511	0.0000	0.2456	0.0989
Keyword in H1/H2/H3 Tag	−0.5404	−0.2359	0.2107	−0.1885
Keyword Density	−0.0677	0.3049	−0.2364	0.0003
Keyword in URL	−0.5890	−0.1362	−0.2443	−0.3232

Table 2. *Cont.*

Seo Factors	Hotel Athens	Email Marketing	Casual Shoes	Average
Keyword in meta description Tag	−0.0323	−0.0691	0.1809	0.0265
Alt text	0.2270	−0.1745	0.1237	0.0587
Title Length	0.0631	−0.3293	0.0000	−0.0887
URL Length	0.4373	0.4908	−0.2716	0.2188
Text to Code ratio	−0.0094	−0.2253	0.0619	−0.0576
Text Length	−0.4582	−0.2330	−0.1727	−0.2880
Quantity of Backlinks	−0.6771	−0.2914	0.0686	−0.3000
404 Page	0.1745	0.0000	0.2474	0.1406
SSL Certificate	−0.3928	−0.3093	−0.3093	−0.3371
Sitemap XML	−0.1260	0.2270	0	0.0337
Number of Site Pages	−0.5652	0.0229	−0.1659	−0.2361
Domain Age	−0.4779	−0.1130	−0.2535	−0.2815
Responsive Layout	0.2474	0.0908	0.0000	0.1127
W3C Validation	0.3712	0.0619	−0.3712	0.0206
Bounce Rate	−0.4942	-0.0092	0.1605	−0.1143
Time on Site	−0.5043	−0.0092	0.3499	−0.0545
Loading Time	−0.6444	0.4415	0.2623	0.0198

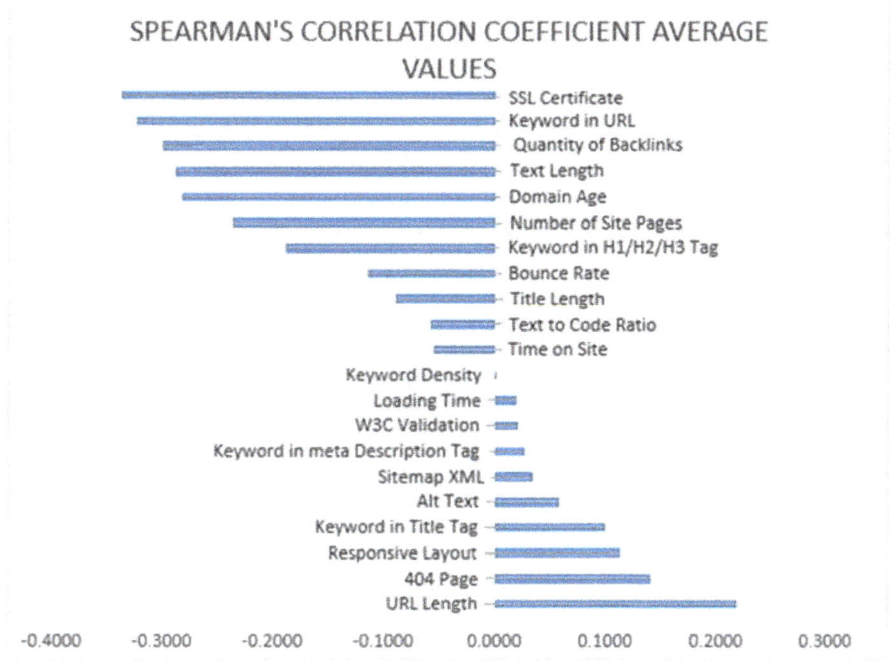

Figure 2. Spearman's correlation coefficient average values.

4.3. Discussion

In comparison with other studies on the topic [26,28,29], we observe that the factors that are crucial despite the changes in algorithms are the quantity and quality of the backlinks and, to a lesser extent, the bounce rate and the SSL certificate. The importance of the quantity is confirmed by our study, but we did not have the means to check whether the quality of backlinks is important too. Contrary to the results of other studies we found out that the website loading time, the URL length, and the use of targeted keywords in the title tag do not affect the ranking of a website, although the difference in the results are due to the approach of our study compared to the others. More specifically, in Dean's study [26], the sample contained less search results for more keywords. In the other one conducted in 2015 by Moz [28], the changes in the algorithm of the search engine may have affected the factors that affect the results.

5. Conclusions

The competition among the websites for the SERPs is huge so a convenient optimization plan is needed: a plan that includes a holistic approach regarding the SEO factors but depends on the most effective ones. As our study revealed, some factors determine the bulk of success in term of ranking high in the SERPs and they have remained unchanged over time. However, search engine algorithms tend to change often, and new factors are added while outdated or not effective factors are excluded. This is why web developers must check the algorithm changes and adjust their websites in order to not only achieve but also maintain high ranking in search engines. This study attempts to show the most effective SEO factors for high ranking in search engines and the significance for the most important factors that affect SEO.

5.1. Implications for Academics

In this paper, 24 SEO factors are included with all of them being equally analyzed, providing the opportunity to explore the most important factors that affect SEO in one place. This study can be used as a base for future studies with the use of better infrastructure in order to reveal more detailed results regarding the factors that affect the ranking of a website in the search engines.

5.2. Practical Implications

Via this research we would like to pinpoint that the results of our study aim not to reveal the patterns and mechanics behind the algorithms of search engines, but to educate and guide the owners of small- and medium-sized businesses, as long as marketers about the most effective factors that optimize the ranking of a website in the search results of a popular search engine such as Google, so that they can be able to plan an effective optimization strategy for their websites without having to use specialized technological tools.

Author Contributions: Conceptualization, C.Z. and M.V.; Methodology, C.Z. and T.K.; Validation, T.K. and M.K.; Formal Analysis, M.K.; Investigation, T.K. and C.Z.; Resources, M.K.; Data Curation, T.K.; Writing—Original Draft Preparation, T.K.; Writing—Review & Editing, M.K.; Visualization, T.K.; Supervision, C.Z.; Project Administration, C.Z. and M.V.

Funding: This research received no external funding.

Conflicts of Interest: The authors declare no conflicts of interest.

References

1. Google Organic CTR History. Fresh CTR Averages Pulled Monthly from Millions of Keywords. 2018. Available online: https://www.advancedwebranking.com/ctrstudy/ (accessed on 25 December 2018).
2. Moher, D.; Liberati, A.; Tetzlaff, J.; Altman, D.G.; the PRISMA Group. Preferred Reporting Items for Systematic Reviews and Meta-Analyses: The PRISMA Statement. *Ann. Intern. Med.* **2009**, *151*, 264–269. [CrossRef] [PubMed]
3. Search Engine Market Share Worldwide—December 2018. Available online: http://gs.statcounter.com/search-engine-market-share#monthly-201806-201806-bar (accessed on 25 December 2018).
4. Dean, B. Backlinko Ranking Factors Study: Methods & Results. 2 September 2016. Available online: https://backlinko.com/wp-content/uploads/2016/02/Search_Engine_Ranking_Factors_Study_Methods__Backlinko.pdf (accessed on 24 January 2018).

5.   Search Engine Ranking Factors 2015. 2015. Available online: https://moz.com/search-ranking-factors (accessed on 6 June 2018).

6.   Ranking Factors SEMrush Study 2.0. 2017. Available online: https://email.semrush.com/acton/attachment/13557/f-0e30/1/-/-/-/-/SEMrush_Ranking_Factors_Study_2_0.pdf (accessed on 6 June 2018).

7.   Hendrickson, B. Statistics a Win for SEO. 2010. Available online: https://moz.com/blog/statistics-a-win-for-seo (accessed on 15 April 2018).

8.   Spearman Rank-Order Correlation Coefficient. Available online: http://vassarstats.net/corr_rank.html (accessed on 21 April 2018).

9.   Yalçın, N.; Köse, U. What is search engine optimization: SEO? *Procedia Soc. Behav. Sci.* **2010**, *9*, 487–493. [CrossRef]

10.   Wilson, R.F.; Pettijohn, J.B. Search engine optimisation: A primer on keyword strategies. *J. Direct Data Digit. Mark. Pract.* **2006**, *8*, 121–133. [CrossRef]

11.   Zilincan, J. Search engine optimization. In Proceedings of the CBU International Conference Proceedings, Prague, Czech Republic, 25–27 March 2015; Volume 3, p. 506.

12.   Cui, M.; Hu, S. Search engine optimization research for website promotion. In Proceedings of the 2011 International Conference on Information Technology, Computer Engineering and Management Sciences (ICM), Nanjing, China, 24–25 September 2011; Volume 4, pp. 100–103.

13.   Rehman, K.; Khan, M.A. The foremost guidelines for achieving higher ranking in search results through search engine optimization. *Int. J. Adv. Sci. Technol.* **2013**, *52*, 101–110.

14.   Wang, F.; Li, Y.; Zhang, Y. An empirical study on the search engine optimization technique and its outcomes. In Proceedings of the 2011 2nd International Conference on Artificial Intelligence, Management Science and Electronic Commerce (AIMSEC), Deng Feng, China, 8–10 August 2011; pp. 2767–2770.

15.   Killoran, J.B. How to use search engine optimization techniques to increase website visibility. *IEEE Trans. Prof. Commun.* **2013**, *56*, 50–66. [CrossRef]

16.   Gupta, S.; Rakesh, N.; Thakral, A.; Chaudhary, D.K. Search engine optimization: Success factors. In Proceedings of the 2016 Fourth International Conference on Parallel, Distributed and Grid Computing (PDGC), Pradesh, India, 22–24 December 2016; pp. 17–21.

17.   Kakkar, A.; Majumdar, R.; Kumar, A. Search engine optimization: A game of page ranking. In Proceedings of the 2015 2nd International Conference on Computing for Sustainable Global Development (INDIACom), New Delhi, India, 11–13 March 2015; pp. 206–210.

18.   Gudivada, V.N.; Rao, D.; Paris, J. Understanding search-engine optimization. *Computer* **2015**, *48*, 43–52. [CrossRef]

19.   Krrabaj, S.; Baxhaku, F.; Sadrijaj, D. Investigating search engine optimization techniques for effective ranking: A case study of an educational site. In Proceedings of the 2017 6th Mediterranean Conference on Embedded Computing (MECO), Bar, Montenegro, 11–15 June 2017; pp. 1–4.

20.   Evans, M.P. Analysing Google rankings through search engine optimization data. *Internet Res.* **2007**, *17*, 21–37. [CrossRef]

21.   Chen, C.Y.; Shih, B.Y.; Chen, Z.; Chen, T.H. The exploration of internet marketing strategy by search engine optimization: A critical review and comparison. *Afr. J. Bus. Manag.* **2011**, *5*, 4644–4649.

22.   Patil Swati, P.; Pawar, B.V.; Patil Ajay, S. Search Engine Optimization: A Study. *Res. J. Comput. Inf. Technol. Sci.* **2013**, *1*, 10–13.

23.   Hui, Z.; Shigang, Q.; Jinhua, L.; Jianli, C. Study on website Search Engine optimization. In Proceedings of the 2012 International Conference on Computer Science & Service System (CSSS), Nanjing, China, 11–13 August 2012; pp. 930–933.

24.   Gregurec, I.; Grd, P. Search Engine Optimization (SEO): Website analysis of selected faculties in Croatia. In Proceedings of the Central European Conference on Information and Intelligent Systems, Varazdin, Croatia, 19–21 September 2012; pp. 211–218.

25.   Chandra, A.; Suaib, M.; Beg, D. Low cost page quality factors to detect web spam. *arXiv*, 2014; arXiv:1410.2085.

26.   Thakur, A.; Sangal, A.L.; Bindra, H. Quantitative measurement and comparison of effects of various search engine optimization parameters on Alexa Traffic Rank. *Int. J. Comput. Appl.* **2011**, *26*, 15–23. [CrossRef]

27.   Egri, G.; Bayrak, C. The role of search engine optimization on keeping the user on the site. *Procedia Comput. Sci.* **2014**, *36*, 335–342. [CrossRef]

28. Kumar, R.; Saini, S. A study on SEO monitoring system based on corporate website development. *Int. J. Comput. Sci. Eng. Inf. Technol.* **2011**, *1*, 42–49.

29. Dean, B. We Analyzed 1 Million Google Search Results. Here's What We Learned about SEO. 2 September 2016. Available online: https://backlinko.com/search-engine-ranking (accessed on 24 January 2018).

30. Fishkin, R. How to Rank in 2018: The SEO Checklist. 2017. Available online: https://moz.com/blog/rank-in-2018 (accessed on 27 December 2018).

31. Palos-Sanchez, P.; Martin-Velicia, F.; Ramon Saura, J. Complexity in the Acceptance of Sustainable Search Engines on the Internet: An Analysis of Unobserved Heterogeneity with FIMIX-PLS. *Complexity* **2018**, *2018*, 6561417. [CrossRef]

 *future internet* 

Article

Academic Excellence, Website Quality, SEO Performance: Is there a Correlation?

Andreas Giannakoulopoulos [1,*], **Nikos Konstantinou** [1], **Dimitris Koutsompolis** [2], **Minas Pergantis** [3,*] and **Iraklis Varlamis** [4,*]

1   Department of Audio and Visual Arts, Ionian University, 7 Tsirigoti Square, 49100 Corfu, Greece; nikoskon@ionio.gr
2   Faculty of Communication and Media Studies, National and Kapodistrian University of Athens, 5 Stadiou Str., 10562 Athens, Greece; dkouts@media.uoa.gr
3   Laboratory of Interactive Arts, Ionian University, 7 Tsirigoti Square, 49100 Corfu, Greece
4   Department of Informatics and Telematics, Harokopio University of Athens, 70 Eleftheriou Venizelou Str., 17676 Athens, Greece
*   Correspondence: agiannak@ionio.gr (A.G.); mperganths@inarts.eu (M.P.); varlamis@hua.gr (I.V.)

Received: 15 October 2019; Accepted: 12 November 2019; Published: 18 November 2019

Abstract: The academic excellence of universities around the globe has always been a matter of extended study and so has the quality of an institution's presence in the World Wide Web. The purpose of this research is to study the extent to which a university's academic excellence is related to the quality of its web presence. In order to achieve this, a method was devised that quantified the website quality and search engine optimization (SEO) performance of the university websites of the top 100 universities in the Academic Ranking of World Universities (ARWU) Shanghai list. A variety of tools was employed to measure and test each website and produced a Web quality ranking, an SEO performance ranking, as well as a combined overall web ranking for each one. Comparing these rankings with the ARWU shows that academic excellence is moderately correlated with website quality, but SEO performance is not. Moreover, the overall web ranking also shows a moderate correlation with ARWU which seems to be positively influenced by website quality and negatively by SEO performance. Conclusively, the results of the research indicate that universities place particular emphasis on issues concerning website quality, while the utilization of SEO does not appear to be of equal importance, indicating possible room for improvement in this area.

Keywords: website quality; search engine optimization; web presence; academic rankings; accessibility

---

1. Introduction

The present research investigates the relevance of universities' performance in international rankings based on academic criteria and their ranking according to the quality of their online presence, their ability to properly promote their content in search engines and finally, in combination, their overall web presence.

Undoubtedly, the idea of a university ranking was not introduced in 2003 with the celebrated Shanghai list or with the TIMES ranking in 2004 [1]. It can be traced as far back as 1900, when there were studies concerning the most successful men in England, linking their professional and broader social achievements with their studies and the universities they had attended. In essence, this was an attempt to correlate the academic quality of universities with indicators less academic and more applicable to everyday life. This correlation was based on measurable and verifiable indicators.

Some might argue that rankings do not always correlate with the actual quality of an institution, but they are a factor for prospective students to decide on whether they will enroll in a specific institute or not [2].

Bell et al. [3] contacted an online survey to examine the scale of online education in Australian universities. These included features such as online courses and access to web libraries. This study had shown that Australian universities had a good level of online components' embracement. In fact, more than 54% of universities were found to contain online features and elements. Since this study is outdated, we can safely assume that not only will this percentage be higher by now, but it will also include more aspects.

In addition, Will et al. [4] performed a study, analyzing more than 3000 university websites in order to examine the ways academic institutions (in higher education) are making use of the Web to facilitate student relations. In this study, the emphasis on higher education websites was given on alumni and their friends and then on prospective students. Will et al. [4] stress how prospective students are going online in order to research academic institutions. McAllister-Spooner [5] confirms that some years later, the web remains the basic research tool for prospective students, counselors and parents.

Baka et al. [2] also affirm that the success or failure of an academic institution to market a program is dependent on its website. When a user and consequently a prospective student—in our case—enters a website, they seek information [6]. In fact, a few years earlier, Abrahamson [7] mentioned that if the potential students do not have a pleasant experience with the website of an academic institution, they will quit the process of enrolling. More than a decade later, Kaur et al. [8] are emphasizing the importance of performance in order for a website to be successful. Kaur et al. [8] analyzed the websites of major universities in Punjab (India) in order to evaluate website elements that could optimize web performance such as speed, SEO and security. Khwaldeh et al. [9], in their study about the relationship between the information quality of e-Services of the websites of Jordanian universities and ICT Competence are mentioning e-Services in general as a basic factor of competence.

In general, there have been many studies that attempt to examine the online presence of academic institutions focusing on usability, user experience or services they provide. The majority of them are referring to universities in a specific location (for instance Jordanian universities, USA universities and Indian universities). What is of great interest is the emphasis given on usability. Conway et al. [10] in a similar study about political websites had argued that usability is more important than quantity. If the information is difficult to be found by a user, then the website will be deemed as less useful.

Furthermore, when it comes to ranking and accessibility and visibility, Baka et al. [2] found that there is a correlation between the top academic institutions on the list, compared to those that they are on the lowest positions on the list. At the same time, Acosta-Vargas et al. [11] on their study on 20 universities found out that despite the ranking and the status of a university, there are still accessibility issues that have yet to be solved, a fact confirmed by Alahmadi et al. [12] who mention that the more websites are evolving, the more information they offer and the more inaccessible to people with disabilities they are becoming.

From the above, it becomes apparent that university websites should have two main goals in mind: Existing students and alumni and potential students. The first goal should be linked with material focused on e-Services and e-Learning while for the second goal, the website should offer informative material. In order for the academic institution's presence to be deemed as satisfactory, the website should keep in mind the terms of usability and accessibility along with material of good quality.

Furthermore, related to our research in the broader sense is the topic of academic SEO (ASEO) that has been gaining attention in the recent years. ASEO applies the SEO principles in the search for academic documents in academic search engines such as Google Scholar and Microsoft Academic. [13]. Towards this, a variety of algorithms are used. All of them utilize the number of citations the articles received combined with other factors like the author's index, the date of publication, the institution etc. [14]. Beel et al. [13] in their study a decade earlier found that most likely Google Scholar and

Microsoft Academic are taking into account the number of citations in order to rank the publications and present them to the user. Just by having a look at these Search Engines today, we can see more or less that the number of citations is still relevant, combined with some other factors, confirming Rovira et al [14].

Beel et al. [13] are stressing the importance for researches to ensure that their publications will receive a high rank on the academic search engines. By making good use of ASEO, researches will have a higher chance of improving the visibility of their publications and have their work read and cited by more researchers. Ale [15] has come up with simple steps that researchers can follow, in order to improve their visibility on the academic search engines. Despite the importance of ranking higher, researchers believe that the authors should not try to hack ASEO in order to approve higher on the list, but use the principles of ASEO in order to help users understand the relevance of the topic [13,14]. In this way, ASEO is similar to traditional SEO, where institutions want to make a good use of a SEO without ending up spamming the users, because this will have an undesirable effect to the end user.

As a measure of academic excellence this research used the Academic Ranking of World Universities (ARWU—also known as the Shanghai list). It is a recognized prestigious university ranking system that is widely accepted and includes measurable and verifiable indicators that reflect the work done in many different aspects of academic life (teaching, research, international relations, and collaborations).

As discussed above, an important parameter for any university these days is the quality of its website presence, i.e., the adequacy of each university's website and its ability to properly present the university's image to the general audience. The work of each university exists only when it encounters and interacts with society. At the same time, this research analyzes the importance of SEO—search engine optimization for universities, as it essentially represents the right concern to facilitate content search; that is to say, it meets to the utmost degree the need of connecting with society.

Therefore, these two key parameters which dominate the digital age, website quality, and SEO (search engine optimization) are combined into a single average index, which is illustrated as web presence, and an attempt is made to investigate their association with the ARWU ranking.

2. International Rankings

International rankings are a tool comparing the multidimensional academic and—mainly—research work undertaken at universities in recent years. Their popularity with the public is steadily growing. Indeed, for the prestigious and traditional higher education institutions, they often constitute a key driver of their growth, as they act as a catalyst in attracting not only new students, but also funders and donors [16].

There is no doubt that there are a large number of higher education institutions operating worldwide. Within this almost chaotic context, it is necessary to establish academic criteria with international common reference and high credibility. Quality assurance systems do not have much history, and efforts at European or international level to cooperate on this issue cannot be said that found common ground.

According to Isidro Aguillo [17], who heads the Webometrics research team, there are 28,000 universities worldwide. Most rankings collect and process data for all of these universities. Based on specific and predefined criteria, they elaborate indicators and the overall score of each university. They then select the first 500 or 600 or 1000 universities and compile the final ranking table. A university in the top 300 with a simple percentage reduction is ranked among the top 1% of the world's best universities. If it is in the top 1000 then it is at the top 3.5% of the best universities in the world. However, we must note that this illustration is an ongoing process, where individual indicators are regularly updated in an effort to make the ranking systems better reflect the true position of the universities.

Ranking is an ever-changing and comparative process. This practically means that, even though a university may go up on individual indicators, if other universities go up faster on the same indicators, that particular university may be lower in the rankings; the same goes for a university that manages to maintain its indicators at the same level. Competition between universities internationally has intensified in recent years and the differences in ranking are in many cases marginal.

Thus, it is necessary to increase the total number of universities included in the ranking lists. For example, from 2003 to 2018 the ARWU ranking, known as the Shanghai list, was publishing rankings of the top 500 universities worldwide. For the first time in 2019 it published a list of the top 1000 universities. Correspondingly, Times Higher Education increased the number of universities included in the rankings from 500 to 800, and the Center of World University Ranking from 1000 to 2000.

2.1. Major Rankings

2.1.1. Academic Ranking of World Universities (ARWU)

The Academic Ranking of World Universities (ARWU) [18], known as the Shanghai list, is perhaps one of the best-known rankings. It was created and is maintained by Shanghai Jiaotong University in China. Ratings have been updated every year since 2003.

The rankings compare and classify the top 1200 higher education Institutions according to an algorithm that includes the following individual indicators:

1. The number of graduates who have received major prizes (Nobel Prize, Fields Medal, etc.).
2. The number of faculty members who have received major prizes (Nobel Prize, Fields Medal etc.).
3. The number of faculty members included in the lists of researchers with a very high number of citations.
4. The number of publications in nature and science journals.
5. The number of publications included in the Science Citation Index and Social Sciences Citation Index databases.
6. The fully measurable per capita academic performance, according to a published algorithm.

The methodology is presented in an article written by the list creators Liu and Cheng [19] in which they state that their original goal was to study "the gap between Chinese and other universities around the world, according to academic criteria".

2.1.2. Webometrics Ranking of World Universities

The Webometrics ranking is compiled by Cybermetrics (CCHS) [20] and it provides information on more than 18,000 universities according to their online presence and the impact of their research work. Universities are ranked by criteria which concern their presence and popularity on the web, the impact of their research work as reflected in the total number of cross-references to the articles and publications of their professors and researchers, as well as the percentage of their publications.

The ranking of each university is based on the following individual indicators:

1. Presence rank;
2. impact rank;
3. openness; and
4. excellence rank

2.1.3. Times Higher Education World University Rankings

Times Higher Education (THE) rankings [21] are considered to be one of the three most influential rankings at international level. In 2009, THE started collaborating with Thomson Reuters to develop a new university ranking system called Times Higher Education World University rankings. The THE World University Rankings methodology includes 13 indicators classified in 5 categories, which are analyzed into individual indicators as follow:

1. Private sector revenue;

2. international dimension;
3. teaching; and
4. research and research impact.

The specific body implementing the evaluation selected 980 universities from an initial selection of 1313 universities that are distinguished in scientific research.

2.1.4. QS World University Rankings

QS World University Rankings [22] include the top 500 universities, were created by Quacquarelli Symonds (QS) and are published annually since 2004. The QS rankings use peer review data collected by researchers, academics, and employers; they also take into account the number of international staff, of students, of Scopus citations and teacher/student ratios.

The QS ranking table for 2017 includes the 950 best universities in the world, after analyzing at least 10,000 Institutions. The scores and rankings in each of the individual five indicators are announced for the top 500 only. The overall score and position in the world rankings are announced for the top 950 universities.

2.1.5. Center for World University Rankings (CWUR)

The Center for World University Rankings (CWUR) [23] publishes a global ranking of universities, perhaps the only one that measures of the quality of the education and training of the student, as well as the prestige of faculty members and the quality of their research without relying on surveys, questionnaires, interviews, and data submission by the universities themselves.

CWUR [23] uses the following indicators, which are based on measurable data and are generally accepted for ranking the top 1000 universities in the world:

1. Quality of education and faculty;
2. alumni employment;
3. publications and citations;
4. influence and wider impact; and
5. patents

2.1.6. US News—Best Global University Rankings

US News rankings of the best universities in the world were created to provide a picture of how universities can compare worldwide [24]. Of all the universities in the world, the final rankings of the 1250 best universities cover a range of 60 countries. The first step in producing these rankings [25] involves the compilation of a first group of top universities, which are used in a second phase to rank the top 1250. For a university to be included in the top 1250 ones, it must first be among the top 250 universities in the results of Thomson Reuters' global research reputation.

The "U.S. News and World Report" ranks the top 1250 institutions using a wide range of different criteria:

1. Global and local research reputation;
2. publications, textbooks, conferences;
3. total and weighted impact of citations;
4. number and percentage of publications that are among the 10% of the most cited publications; and
5. international collaborations.

2.1.7. National Taiwan University (Performance Ranking of Scientific Papers)

The ratings of National Taiwan University are mainly based on the research work produced and published by each Institution, on the impact of the published work, and on their final particular distinction [26]. They are under the auspices of the Taiwan Higher Education Quality Assurance Authority and have been ranking the top 500 universities in the world since 2007, based solely on their scientific publications and the impact they have on the international scientific community.

The key axes of evaluation are summarized in three basic variables: Research production, research impact, and research excellence.

The main points of all the above are summarized in Table 1:

Table 1. Summary of the main characteristics of the major rankings.

Subsection	Ranking Name	Main Indicators/Criteria	Generic Characteristics
Section 2.1.1	Academic ranking of world universities (ARWU)/Shanghai list	The number of graduates and faculty members who have received major prizes. The number of faculty members included in the lists of researchers with a very high number of citations. The number of publications in nature/science journals and in the science/ social sciences citation index. The normalized per capita academic performance.	One of the best-known rankings criteria—academic and oriented.
Section 2.1.2	Webometrics ranking of World universities	Presence rank. Impact rank. Openness. Excellence rank.	Criteria which concern their presence and popularity on the web and the impact of their research work.
Section 2.1.3	Times Higher Education World University Rankings	Private sector revenue. International dimension. Teaching. Research and research impact.	One of the most influential rankings at international level
Section 2.1.4	QS World University Rankings	Peer review data. The number of international staff, students, Scopus citations and teacher/student ratios.	The scores and rankings in each of the indicators are announced for the top 500 only.
Section 2.1.5	Center for World University Rankings (CWUR)	Quality of education and faculty. Alumni employment. Publications and citations. Influence and wider impact. Patents.	Measures of the quality of the education and training of the student, the prestige of faculty members and the quality of their research.
Section 2.1.6	US News—Best Global University Rankings	Global and local research reputation. Publications, textbooks, and conferences. Total and weighted impact of citations. Number and percentage of publications that are among the 10% of the most cited publications. International collaborations.	Were created to provide a picture of how universities can compare worldwide [18].
Section 2.1.7	National Taiwan University (Performance ranking of scientific papers)	Research production. Research impact. Research excellence.	Based on the research work produced and published by each institution, on the impact of the published papers, and on their final particular distinction.

2.2. Choosing the Most Appropriate Academic Excellence Ranking

The selection of the Academic Ranking of World Universities (ARWU), known as the Shanghai list [18], is ultimately the most consistent choice for evaluation rankings as it combines the following characteristics:

1. It is one of the most popular rankings, with many years of publications and consistent credibility over the years.
2. All of the indicators used to create the ranking are measurable.
3. All of the indicators used to create the ranking are publicly verifiable by third parties.

4. Evaluations are updated every year. Since 2009 the rankings have been published by the Shanghai Ranking Consultancy.

As it has already been mentioned, the ranking table includes various indicators. For each indicator, the highest-performing institution receives a score of 100, so all other institutions receive a score below the score of the first one. The indicators are [27]:

1. The number of graduates who have received major prizes (Nobel Prize, Fields Medal, etc.).
2. The number of faculty members who have received major prizes (Nobel Prize, Fields Medal etc.).
3. The number of faculty members included in the lists of researchers with a very high number of citations.
4. The number of publications in nature and science journals.
5. The number of publications included in the Science Citation Index and Social Sciences Citation Index databases.
6. The fully measurable per capita academic performance, according to a published algorithm.

The criteria of ARWU are therefore purely academic and largely focused on research. The research characterizes universal universities and distinguishes them from colleges and education centers. Moreover, the Shanghai list publishes rankings based on data that anyone can verify. It draws on data from official authorities in each country and avoids both data that may be questionable and subjective evaluation judgments, such as opinion polls, which, although based on common sense, actually raise questions about the way they are carried out and the weight assigned to them in many rankings (in 30–40% of them). Finally, it is important to use scaled scores for each ranking, but each individual score must always be relative to the highest score in its ranking, i.e., on a 100-degree scale, the first institution receives a score of 100 and all the others receive a score below that.

Additionally, the ranking system has not remained static over the course of its existence. In an effort to measure achievements in multiple different fields of study while retaining the same transparency of its criteria the Shanghai ranking has engaged with over 6000 professors from top 100 universities globally [28]. These individuals whose names are publicly available have been surveyed and their answers were used to create a list of internationally accepted awards and journals which are recognized as top awards or top journals and which are used in the ranking. Even now, any professor from a top 100 institution can actively contribute to this survey.

Despite the fact that other rankings focus more on the teaching aspect or the overall presence of an institution (including its online presence) the result focused nature of Shanghai was more suited as a starting point for our research. Evaluation methods on these different aspects can be a source of subjectivity while no one can argue against the output-oriented methodology of the ARWU.

For this reason, the Shanghai list is the most reliable and appropriate one for the purposes of the present research and, therefore it is the one that was selected.

3. Methodology

With regards to methodology, the research process consisted of a series of exact measurements and tests that were carried through using a set of reckoning tools specifically developed for their appropriate purpose. All measurements and tests took place in September 2019. In order to consolidate the different measurements and test results into ratings, the gravity of each measurement was judged by its scope and importance and the measurement was assigned a weight according to its contribution in evaluating the equivalent characteristic of each website. Then each rating was calculated by means of a weighted average. The rationale behind each weight is touched upon but ultimately the evaluating process is a result of the researchers' experience with both the reckoning tools and the specific requirements of website design and development.

3.1. Measuring the Quality of a Website

In order to examine whether there exists a quantifiable correlation between a university's academic excellence ranking and the quality of its website it was necessary to quantify the latter through a series of tests and measurements. The results of these tests and measurements were then used in a unifying equation that yielded a metric of website quality in a 100-degree scale.

For the purpose of this quantification we assumed three major characteristics of a website that indicate its quality:

- Website structure;
- website accessibility; and
- website performance.

For each of these characteristics multiple tests were performed and the results of these tests were combined to achieve a rating that was an accurate representation of the characteristic.

3.1.1. Evaluation of Website Structure

The structure of each university's website was evaluated based on the following aspects:

- Validity of HTML;
- validity of CSS;
- Google's mobile friendliness test; and
- Google lighthouse best practices audit.

Evaluation of a Website's HTML

One of the most important aspects of a website's structure is the validity of its HTML. A valid code ensures a website's compatibility with multiple browsers, the uniformity and speed of rendering as well as a sound foundation for future technology support. It also positively affects both SEO performance and accessibility, both metrics that were measured individually for the purpose of this methodology [29].

In order to test HTML validity, the Markup Validation Service of the World Wide Web Consortium (W3C) [30] was employed. W3C is "an international community where different member organizations, a full-time staff, and the public work together to develop Web standards" [31]. It is the most prominent international web standards organization and is widely acknowledged by the scientific community. "Most Web documents are written using markup languages, such as HTML or XHTML. These languages are defined by technical specifications, which usually include a machine-readable formal grammar (and vocabulary). The act of checking a document against these constraints is called validation and this is what the Markup Validator does" [32]. The W3C Validator also conforms "to International Standard ISO/IEC 15445—HyperText Markup Language, and International Standard ISO 8879—Standard Generalized Markup Language (SGML)" [32]. For the purposes of this test the homepage of every university website was passed through the W3C validator. This action yielded as a result the number of HTML errors and warnings present in each website's HTML. The websites were graded based on errors by subtracting one point off their HTML error score on a 100-degree scale for each error (as long as the score remained above or equal to zero). Similarly, by subtracting one point off their HTML Warning score for every 10 warnings they were graded on warnings. The increased number of warnings required for the subtraction of a point was decided because warnings are both much more common and also much less detrimental to the overall quality of the HTML.

Evaluation of a Website's CSS

A website's appearance is not only a result of its HTML code but also of the CSS sheet embedded in the HTML document. All modern websites make ample use of CSS's inherent ability to improve their appearance, but an invalid CSS can cause, compatibility, usability, and performance issues [33]. In a similar way to the HTML process, W3C's CSS Validation Service [34] was used to test the websites' validity. "The validity of a style sheet depends on the level of CSS used for the style sheet. [...] valid CSS 2.1 style sheet must be written according to the grammar of CSS 2.1. Furthermore, it must contain only at-rules, property names and property values defined in this specification" [35]. The grading method was identical to the one used for HTML in order to achieve consistency of the results. This means for every error one point was deducted from the error score on a 100-degree scale and for 10 warnings one point was deducted from the warning score.

Mobile Friendliness

In the modern world, website access through mobile devices has overtaken desktop access [36], which makes the mobile friendliness of a website an extremely important factor. The websites' ability to display properly on mobile was tested through Google's appropriate test [37]. Google has been at the forefront of the World Wide Web innovation for more than two decades and provides multiple impartial and very reliable tools to evaluate all aspects of a website. Many of these tools have been used throughout this methodology. The mobile friendliness test checks for a series of errors that might occur in a page when displayed on a mobile device. These errors are: "Use of incompatible plug-ins, viewport not set, viewport not set to "device-width", content that exceeds the screen width, text too small to read and clickable elements too close together" [38]. For the purpose of evaluating a Website's structure, mobile friendliness was treated as a binary aspect. Websites that were mobile friendly received full marks in a 100-degree scale, while websites that were not, received none. Other aspects of mobile usability such as performance were measured in the later parts of the methodology.

Best Practices

Besides validation and mobile compatibility there are some other aspects of a website that indicate its effort for maximum compatibility and usability. Google Lighthouse's best practices test measures a series of such aspects [39]. "Google Lighthouse is an open-source, automated tool for improving the quality of web pages" [40] that can be used through Google Chrome's development tools. This tool has been used in multiple instances of our testing methodology as it combines Google's reliability with ease of use and access. Specifically the best practices test check that a website performs the following: "Avoids Application Cache, Avoids console.time(), Avoids Date.now(), Avoids Deprecated APIs, Avoids document.write(), Avoids Mutation Events In Its Own Scripts, Avoids Old CSS Flexbox, Avoids Requesting The Geolocation Permission On Page Load, Avoids Requesting The Notification Permission On Page Load, Displays Images With Incorrect Aspect Ratio, Includes Front-End JavaScript Libraries With Known Security Vulnerabilities, Manifest's short_Name Won't Be Truncated When Displayed On Homescreen, Links to cross-origin destinations are unsafe, Prevents Users From Pasting Into Password Fields, Some Insecure Resources Can Be Upgraded To HTTPS, Uses HTTPS, Uses HTTP/2 For Its Own Resources, Uses Passive Event Listeners To Improve Scrolling Performance, Avoids Web SQL" [39].

Consolidating Website Structure Ratings

The different aspects that were discussed above were consolidated in a single Website Structure Evaluation metric. This single metric was calculated as a weighted average of all individual metrics. The purpose of assigning weights was to finetune the influence of each metric on the final individual rating according to its importance as judged by the researchers. Since both HTML and CSS errors were deemed much more damaging to a website than warnings, both error metrics received a higher weight, while warning metrics received a much lower one. This is also in accordance with the rate of appearance of each: Warnings are much more common. Mobile friendliness should be considered of utmost importance in a continuously more mobile environment so it received a weight that sets it in the same level of importance as HTML or CSS validity as a whole. Google Lighthouse's best practices audit is used to check for secondary practices that improve a website's quality, so it is valued a little lower than the other major metrics. The weight of each individual metric contributing to the consolidated Website Structure Evaluation metric is presented in Table 2.

Table 2. Weights for the calculation of the website structure evaluation metric.

Metric	Weight
HTML errors	22
HTML warnings	5.5
CSS errors	22
CSS warnings	5.5
Mobile friendliness	27
Lighthouse best practices	18

In addition to the initial scores a bonus was given for websites that had a perfect error score in either the HTML or CSS error metrics. While having as few errors as possible is always a good sign of the care and attention given to a website, having no errors whatsoever indicates an adherence to practices that is even more commendable. Because of that, websites that took the extra mile to achieve zero errors were awarded with 2 extra points on a 100-degree scale in their final scores (to a maximum of 4 points if both error ratings were flawless). In order to avoid ruining the scale's proportionality the final result was recalculated as a percentage of the maximum score that was 104.

3.1.2. Evaluation of Website Accessibility

The accessibility of each university's website was evaluated based on the following aspects:

- WCAG 2.0 compatibility problems as indicated by aChecker;
- WCAG 2.0 compatibility problems as indicated by the WAVE accessibility tool; and
- Google Lighthouse accessibility audit.

Evaluation of a Website's WCAG 2.0 Compatibility

Accessibility is an issue of extreme importance in all aspects of human activity and the World Wide Web is no exception [41]. It is a website's primary objective to be inclusive of the maximum number of individuals and that is especially true for websites of academic nature. The Web Content Accessibility Guidelines (WCAG) "are developed through the W3C process in cooperation with individuals and organizations around the world, with a goal of providing a single shared standard for web content accessibility that meets the needs of individuals, organizations, and governments internationally" [42]. A website's adherence to these guidelines is a good indicator of the care given to make the website accessible and usable. Because the nature of these guidelines is not as strict or well-defined as the syntax of HTML or CSS, in order to get a more accurate measurement, two distinct WCAG validators were used. The first was aChecker [43], "a Web accessibility evaluation tool designed to help Web content developers and Web application developers ensure their Web content is accessible to everyone regardless of the technology they may be using, or their abilities or disabilities" [43]. The second was WAVE [44], a tool provided by WebAim, "an organization dedicated to bringing more accessibility to the Web" [45]. Both of these tools are mentioned and linked in W3C's official Web Accessibility Evaluation Tools List [46].

For the purpose of the test, the frontpage of every university's website was processed through both tools and four individual metrics were established. One metric for aChecker's known problems, one for aChecker's potential problems, one for WAVE's Errors and one for WAVE's alerts. In a manner equivalent to our process during the website structure evaluation one point was subtracted for every aChecker known problem, WAVE error or WAVE alert from a 100-degree scale and one for every ten aChecker potential problems. The reason potential problems are measured by tens is because they were much more numerous on average.

Evaluation of a Website's Accessibility with Google Lighthouse

Google's Lighthouse was used to get a complimentary Accessibility metric. Lighthouse measures multiple accessibility aspects and returns a weighted average in a 100-degree scale. The tests run to determine accessibility check for the following: "Buttons Have An Accessible Name, Document Doesn't Have A Title Element, Every Form Element Has A Label, Every Image Has An alt Attribute, No Element Has A tabindex Attribute Greater Than 0" [39].

Consolidating Website Accessibility Ratings

The different aspects that were discussed above were consolidated in a single Website Accessibility Evaluation metric. This single metric was calculated as a weighted average of all individual metrics. aChecker known problems and WAVE errors were deemed more important than potential problems and Alerts so they received a higher weight. Potential problems and alerts received a lower weight. This is also in accordance with the rate of appearance of each, similarly to the HTML/CSS errors and warnings that were measured in the structure section. The Google Lighthouse accessibility audit's rating was given an overall lower weight than the major metrics since its scope is limited and tends to provide a more generalized estimate of accessibility. The weight of each individual metric contributing to the consolidated Website Accessibility Evaluation metric is presented in Table 3.

Table 3. Weights for the calculation of the website accessibility evaluation metric.

Metric	Weight
aChecker known problems	30
aChecker potential problems	7.5
WAVE errors	30
WAVE alerts	7.5
Lighthouse accessibility	25

In addition to the weights, a bonus system was employed similar to the one used on the website structure evaluation metric. Perfect scores on aChecker known problems and WAVE errors were rewarded with 2 points each up to a maximum of 4 and the final result was recalculated as a percentage of the maximum score.

3.1.3. Evaluation of Website Performance

A website's performance is quintessential to its presence in the World Wide Web. It helps it retain the attention of users and increase each user's level of engagement [47]. It streamlines user experience and the overall quality of our virtual lives.

The performance of each university's website was evaluated based on the following aspects:

- WebPagetest's performance metrics
- Google PageSpeed Insights for mobile and desktop

Evaluation of a Website's Performance with WebPagetest

WebPagetest [48] is an open source tool for measuring different aspects of a website's performance. Its online version, which was used for the purposes of this methodology, "is hosted by multiple companies and individuals providing the testing infrastructure around the globe" [49]. It is an efficient and impartial tool that has been used for more than a decade and is widely trusted by the online community.

For each university's website, the first view of its frontpage was tested and measured. The tests were run using Chrome and form a trusted location near the university's physical location. These locations included Dulles VA USA, Ireland EU, Sydney Australia, Seoul Korea, and Tokyo Japan. The

USA tests were carried out using WebPagetest's own infrastructure while all the rest were carried out from infrastructure provided by Akamai through EC2 in order to achieve maximum conformity.

The metrics taken into consideration were:

- First Byte

  "The First Byte time (often abbreviated as TTFB) is measured as the time from the start of the initial navigation until the first byte of the base page is received by the browser (after following redirects)" [50].

- Usage of the Keep-Alive extension

  "The Keep-Alive extension to HTTP/1.0 and the persistent connection feature of HTTP/1.1 provide long-lived HTTP sessions which allow multiple requests to be sent over the same TCP connection. In some cases, this has been shown to result in an almost 50% speedup in latency times for HTML documents with many images" [51].

- Compression of documents

  "Compression is an important way to increase the performance of a Web site. For some documents, size reduction of up to 70% lowers the bandwidth capacity needs" [52].

- Compression of images

  "Images often account for most of the downloaded bytes on a web page and also often occupy a significant amount of visual space. As a result, optimizing images can often yield some of the largest byte savings and performance improvements for your website" [53].

- Use of progressive images

  Progressive images are images that display first a lower quality of themselves in full dimensions and then achieve a higher quality by the end of their load time. It is a smart way to optimize a user's experience and reduce bandwidth [54].

- Cache Static Content

  "Static Content are the pieces of content on your page that don't change frequently (images, javascript, css). You can configure them so that the user's browser will store them in a cache so if the user comes back to the page (or visits another page that uses the same file) they can just use the copy they already have instead of requesting the file from the web server" [55].

- Use of a Content Delivery Network (CDN).

Websites use networks of proxy servers and data servers known as CDNs to serve their static data to users without having to directly communicate with the original website server. This enables each user to be served by their closest node thus improving performance by decreasing the data's travel time.

WebPagetest's First Byte rating is provided in a scale from A to F. In order to convert the First Byte rating into a 100-degree scale, numerical values were assigned to different grades: A = 100, B = 80, C = 60, D = 40, and F = 20. The rest of the values were provided by WebPagetest as percentages.

Evaluation of a Website's Performance with Google PageSpeed Insights

While WebPagetest provides insight strictly to a website's network performance values, Google PageSpeed utilizes not only lab results and metrics provided by Google Lighthouse, but also data provided by the Chrome user experience report, which includes both First Contentful Paint and First Input Delay metric data [56]. This means that previous user experiences with how a website performed were also taken into account. This led to the decision to include Google PageSpeed Insights' data in our test methodology. PageSpeed's final result is a separate grade in the 100-degree scale for mobile and desktop views of the website. In order to measure the mobile version's performance, the system emulates the performance of a mid-tier mobile device on a mobile network [57].

Consolidating Website Performance Ratings

The different aspects that were discussed above were consolidated in a single Website Performance Evaluation metric. This single metric was calculated as a weighted average of all individual metrics. WebPage tests First Byte, transfer compression, image compression and static cache measurements all receive the same weight as they are considered more or less equally important to achieve good performance from a technical perspective. Keep-alive functionality was given a lower weight since it is a very common feature for almost all modern websites. Progressive images were given an even lesser weight since their use, or lack thereof, may not always be beneficial especially with specific image types. Each website's CDN usage received double the weight of the average WebPageTest metric not only because it is considered especially important for the websites of world-renowned Institutions, but also due to its contribution to their overall global presence. In order to keep WebPageTest's influence of the final result in line with Google PageSpeed insights' both the desktop and mobile measurements of performance by Google were given the appropriate weight to bring Google's measurements to an almost equal footing since both systems are trustworthy and widely used. The slight edge goes to WebPageTest mainly because of its CDN measurement. The weight of each individual metric contributing to the consolidated Website Performance Evaluation metric is presented in Table 4.

Table 4. Weights for the calculation of the website performance evaluation metric.

Metric	Weight
WPT first byte	7
WPT transfer compression	7
WPT Img compression	7
WPT static cache	7
WPT keep-alive	4
WPT progressive	3
WPT CDN	13
Google PageSpeed mobile	26
Google PageSpeed desktop	26

Consolidating All Ratings into a Comprehensive Website Quality Rating

The values received through rigorous and meticulous measuring and testing were consolidated into ratings for all three of the major assumed characteristics of a website's quality (structure, accessibility, and performance). In order to reach an all-encompassing rating these three ratings were unified as a weighted average. The purpose of assigning weights was to finetune the influence of each metric on the final overall rating according to its importance as judged by the researchers.

Each website's structure received a standard weight value as it represents the most basic of the three characteristics: An essentiality that every website must adhere to.

Each website's accessibility received a higher weight value since it represents a quality that is paramount due to the nature of both the Web and education. The Web is an intrinsic part of our everyday lives and as such its availability to people with disabilities is of utmost importance and a serious obligation. Similarly, ensuring accessibility for higher education has been an ongoing effort for decades and has been an incentive not only for affirmative action but also for legislation reform in many parts of the world.

Each Website's performance received a weight value between those of structure and accessibility. Performance is important in maintaining a user's attention and increasing both retention and engagement in a modern website.

Using these three values and their equivalent weights we calculated a rating indicative of the quality of each website. The weights of the values used in the calculation of the website quality rating are presented in Table 5.

Table 5. Weights for the calculation of the overall website quality evaluation metric.

Metric	Weight
Website Structure	27
Website accessibility	40
Website performance	33

3.2. Measuring a Website's SEO Performance

Search Engine Optimization is a series of methods used to increase the traffic of a website through attaining better ranking results in search engines and thus gaining better visibility in the Web as a whole. "Generally, the earlier, and more frequently a site appears in the search engine results page, the more visitors it will receive from the search traffic. In other words, it is a set of techniques that take into account the evaluation criteria of search engines regarding website content and structure" [58]. Although not part of a website's quality in the strict sense of the term it is a good indicator of the effort put into making the website available to a larger audience. As such it is important to also get a clear picture of the relationship between a university's academic excellence ranking and its website's SEO performance.

Measuring a website's SEO performance is not an exact science since a lot of the mechanics behind gaining rank in Search Engine rankings are under wraps. Despite that, there are multiple tools that use available guidelines, provided by Search Engine developers such as Google, as well as other methods to quantify a website's achievements in this aspect. These tools are usually commercial and as such they do not disclose their exact methodology, techniques or practices. This makes the results opaque and maybe even questionable. In order to get a more impartial rating for each university's website multiple tools were used. On top of that, a method was devised that was employed to normalize extreme differences in the results. It is further explained later.

The tools used for this metric were:

- Neil Patel's SEO analyzer;
- SEO site checkup;
- WooRank website review; and
- Google Lighthouse SEO audit

The Web marketing dedicated website Neil Patel [59] has been a staple in the SEO research community for the past few years and has been recently updated. It uses a crawling algorithm that scans through each website's 150 first pages for common SEO related errors and quantifies the result [60]. Besides SEO errors it also focuses on the SEO quality of the content, which is measured by keyword usage, number of words per page, size of page titles, and descriptions and similar metrics.

The website SEO Site Checkup [61] is a commercial tool used to measure the SEO efficiency of competitors and provide insight for the improvement of a website's SEO optimization [62]. Alongside the usual metrics the tool provides an array of Advanced SEO such as the use of HTML Microdata specifications or the existence of a dedicated 404 page.

WooRank [63] is a tool that boasts more than one million registered users. It uses a variety of methods to measure the overall SEO performance of a website including usage of backlinks, mobile optimization, and social media connectivity [64]. Mobile friendliness is not only measured by performance but also secondary metrics such as the size of tap targets etc. The emphasis on social media is measured both by evaluating engagement in the website's linked social media pages and also by keeping track of how many times a website has been shared in the major social media networks.

Google Lighthouse SEO audit tests for adherence to the essential structure of SEO related data that is provided in Google's own guidelines. The test checks for the following errors: "Document Does Not Have A Meta Description, Document Doesn't Have A Title Element, Document doesn't have a valid hreflang, Document doesn't have a valid rel=canonical, Document Doesn't Use Legible Font Sizes,

Document uses plugins, Links Do Not Have Descriptive Text, Page has unsuccessful HTTP status code, Page is blocked from indexing, robots.txt is not valid, Tap targets are not sized appropriately" [40]. All of the SEO tools employed in this study are listed in Table 6.

Table 6. Search engine optimization (SEO) tools overview.

Tool	Company	URL	Focus
Neil Patel's SEO analyzer	Neil Patel digital	neilpatel.com	Crawling multiple pages, content metrics (word counts, title sizes etc.).
SEO site checkup	SEO site Checkup LLC	seositecheckup.com	Advanced SEO tests, competition comparisons.
WooRank website review	WooRank	www.woorank.com	Mobile friendliness, social media.
Lighthouse SEO audit	Google LLC	developers.google.com/web/tools/lighthouse	Essential structure, Google guidelines.

Calculation of a single metric for a site's SEO performance was achieved through means of a weighted average. In general, the different SEO tools all provided us with a rating in the 100-degree scale and to a large extent preformed similar tests that they evaluated differently.

Neil Patel's results received a standard weight. The crawling method it uses provides a wider sample for testing but the tool's content-oriented testing seems to have shortcomings when dealing with academic websites. This is demonstrated by a lack of variation in ratings and a small standard deviation.

SEO Site Checkup received a slightly higher weight. As a commercial application it provides a vast variety of different checks and that results in a bigger standard deviation of result values. Its performance and measurements were deemed the most trustworthy.

WooRank received a standard weight. Its main contribution beyond basic SEO structure and usage metrics is its focus on mobile friendliness and social media. Both these aspects are becoming ever more important in the world wide web.

Finally, Google Lighthouse's SEO audit covered the most basic prerequisites and so received a slightly lower weight.

The weight of each individual metric contributing to the consolidated SEO Performance Evaluation metric is presented in Table 7.

Table 7. Weights for the calculation of the SEO performance evaluation metric.

Metric	Weight
Neil Patel	25
SEO site checkup	30
WooRank	25
Lighthouse SEO	20

Since all four methods yield a result in the 100-degree scale and all use similar methods to calculate SEO performance, it was deemed necessary, in order to ensure the impartiality of the results, to diminish the weight of measurements that diverge from the norm. To achieve that, our algorithm checks if a rating given by any of the tools is further away than the equivalent average distance between itself and all of the other three tools. This way an outlier value is identified. If that is the case the weight of this specific tool for this specific university website's rating is halved.

For example, if a website got ratings of 75, 72, 32 and 80 from the different tools, the weight of the tool that returned the outlier result (32) would have been halved in order to avoid the outlier value bearing too much influence.

This also holds true if more than one outlier is discovered. If all the tools give results that are above the appropriate average distances, then all weights are halved except for SEO site checkup's weight that has been deemed as the most trustworthy tool.

In the end we found ourselves with a single value indicating a website's overall SEO performance on a 100-degree scale.

3.3. Web Presence as a Combination of Quality and SEO Performance

Even though website quality and SEO performance are different metrics they both represent, in tandem, a more abstract notion of a website's strength that can be defined as the website's overall web rating. Assigning a value to this is as simple as taking the average of both the major metrics as described above (Web quality and SEO performance). This value's relation with a university's academic excellence ranking was also studied.

The three major scales (website quality rating, SEO performance rating and overall web rating) were used to create the three individual rankings that we used in our calculations: Web quality ranking (WQR), SEO performance ranking (SEOPR), and overall web ranking (OWR).

4. Results

Table 8 depicts our data formation. The first column shows the top ten universities according to their ranking on a 100-degree scale, the second column presents institution ranking according to ARWU. Columns 3, 4, and 5 show the corresponding ratings of our measurements on a 100-degree scale, website quality rating, SEO performance rating, and overall web rating. Columns 6, 7, and 8 show the corresponding rankings namely website quality ranking (WQR), SEO performance ranking (SEOPR) and overall web ranking (OWR). The whole table with 100 universities can be found in the dataset provided in Supplementary Materials.

Table 8. Data formation.

Institution	ARWU Rating	ARWU Ranking	Website Quality Rating	SEO Performance Rating	Overall Web Rating	WQR	SEOPR	OWR
Harvard U.	100.00	1	80.67	88.46	84.56	21	14	10
Stanford U.	75.10	2	78.92	87.71	83.32	35	18	17
U. Cambridge	72.30	3	79.51	86.49	83.00	30	26	20
MIT	69.00	4	79.39	81.96	80.67	31	63	36
UC Berkeley	67.90	5	80.23	83.17	81.70	24	57	34
Princeton U.	60.00	6	79.59	84.05	81.82	28	45	32
U. Oxford	59.70	7	80.53	84.79	82.66	23	36	21
Columbia U.	59.10	8	68.12	80.24	74.18	71	75	72
California Tech	58.60	9	48.26	79.40	63.83	97	78	95
U. Chicago	55.10	10	79.18	89.38	84.28	33	8	13

To analyze the interrelation between ARWU rankings and our WQR, SEOPR, and OWR, Spearman rho coefficients have been applied due to the ordinal scale of data [65]. The results are shown in Table 9 where there appears to be a correlation between Shanghai's ranking and WQR (0.355) and OWR (0.32), which leads us to reject the null hypothesis (that there might be no correlation between ARWU and our rankings). The significance level is 0.000 and 0.001 respectively confirming the statistically significant correlation. However, there is no correlation between ARWU and SEOPR, which leads us to conclude that the SEO may not be particularly taken into account by major academic institutions. Further analysis may verify the first results of the survey.

Table 9. Spearman correlation coefficient between the Academic Ranking of World Universities (ARWU)'s 100 highest universities ranking and our website quality ranking (WQR), SEO performance ranking (SEOPR), and overall web ranking (OWR).

	WQR	SEOPR	OWR
Shanghai Ranking	Rho (100) = 0.35 (Sig. 2 sided 0.000)	Rho (100) = −0.044 (Sig. 2-sided 0.333)	Rho (100) = 0.32 (Sig. 2-sided 0.001)

The 100-degree scale on both the ARWU ratings and our ratings allows us to go a step further in the statistical criterion of regression analysis in order to measure to what extend there is a correlation between our variables [66,67]. Since there is no correlation between ARWU and SEOPR, and variable OWR is derived from the mean of WQR and SEOPR, there is no reason to use multiple regression analysis but simple regression between ARWU rating and website quality rating. In our case the independent variable is ARWU rating and the dependent variable is website quality rating.

First, we need to calculate the Pearson correlation coefficient to see if there is a correlation between them. The Pearson correlation is 0.266 at 0.007 2-sided significance, which shows a weak but statistically significant result.

At the model summary in Table 10, we can observe that according to R^2 the 7% of the total variance of website quality is due to ARWU with standard estimation error of 10.35 which is a slightly weak outcome.

Table 10. Model summary of simple regression analysis.

Model	R	R Square	Adjusted R Square	Std. Error of the Estimate
1	.266 [a]	.071	.062	10.35

a. Predictors: (Constant), Shanghai rating.

The variance analysis in Table 11, examines the null hypothesis that there is no linear relationship between website quality and ARWU. The result F (1.98) = 7.04, p = 0.007 < 0.05 is statistically significant and we reject the null hypothesis. Therefore, our predictive model is considered sufficient to continue the additional evaluation of simple regression. In the following Table 11 it is depicted that the value of B is 0.224 (0.007 < 0.01). We also find that there is a positive correlation between the two variables, that is, for each unit increasing ARWU we have an increase of 0.22 units of website quality.

Table 11. Model coefficients simple regression analysis.

Model		Unstandardized Coefficients		t	Sig.
		B	Std. Error		
1	(Constant)	63.6	3.2	19.87	.000
	Shanghai rating	.224	.082	2.737	.007

The regression analysis test between ARWU and website quality is depitcted in the histogram and scatterplot of Figure 1. The histogram shows that the slightly skewed, distribution satisfies the normality assumption. The scatterplot has the predicted values in the x-axis and residuals on the y-axis as shown below. Although it shows a slight "heteroscedasticity" (meaning that the residuals get larger as the prediction moves from large to small), the regression coefficients still have a correlation. However, we cannot use this model as a strong and accurate predictive approach.

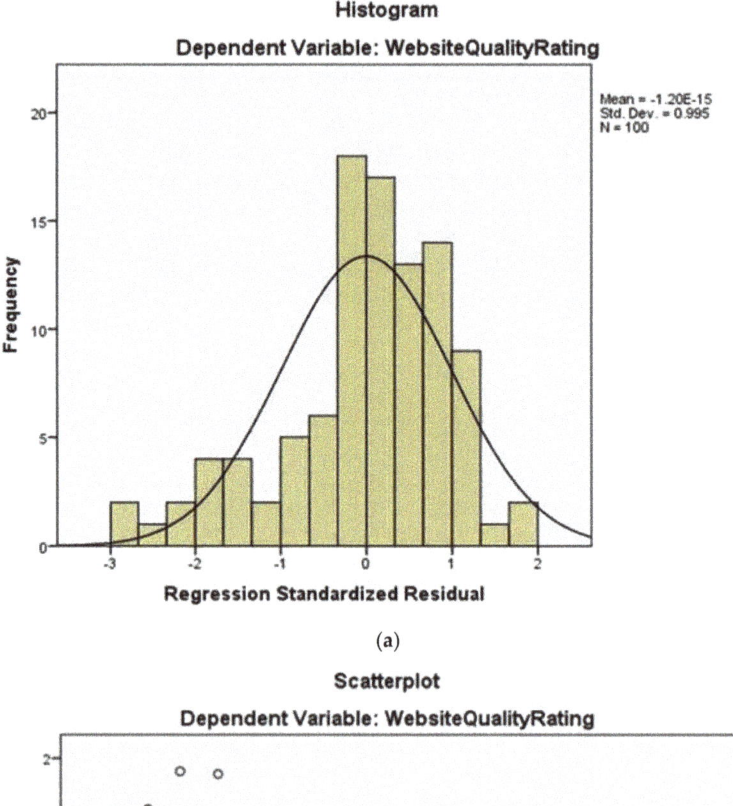

(a)

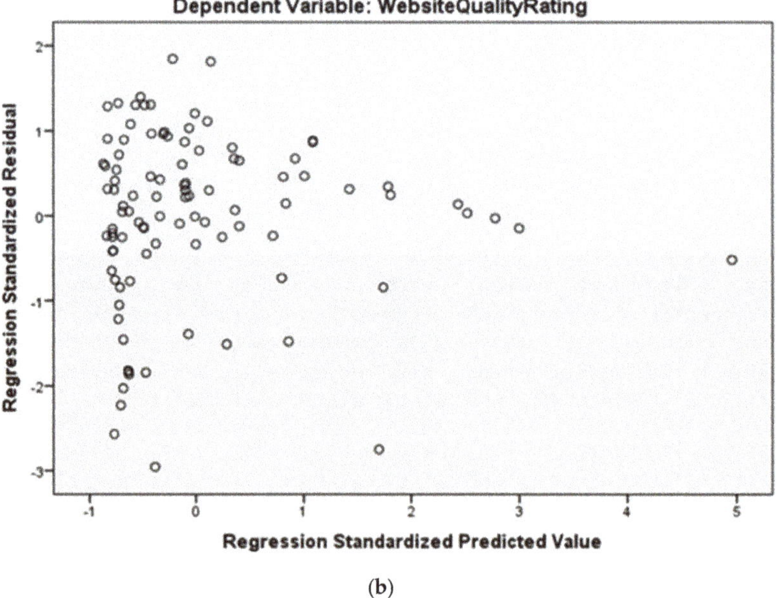

(b)

Figure 1. (**a**) The histogram of the standardized residuals shows a normal distribution. (**b**) The scatterplot shows that there are no clear patterns except a slight "heteroscedasticity" which makes this predictive model uncertain.

To investigate the influence of weights on the outcome of the relationship between Shanghai ranking and WQR, SEOPR, and OWR, we repeated the test using flat weights. Table 12 shows that the Spearman correlation coefficient decreases for all variables WQR, SEOPR, an OWR in a similar way. This proves that the correlation between Shanghai ranking, WQR and QWR exists and the use of weights makes this relationship more explicit.

Table 12. The use of flat weights in the calculation of Spearman correlation coefficient between the ARWU, website quality ranking (WQR), SEO performance ranking (SEOPR), and overall web ranking (OWR).

	WQR	SEOPR	OWR
Shanghai ranking	Rho (100) = 0.31 (Sig. 2-sided 0.002)	Rho (100) = −0.06 (Sig. 2-sided 0.55)	Rho (100) = 0.28 (Sig. 2-sided 0.004)

We found that ARWU ratings and website quality are correlated. However, investigating the correlation between website quality and SEO performance can reveal further information. The aim of this step is to examine these variables' association in order to draw useful conclusions about their co-variation. The simple regression follows by setting website quality as the independent variable and SEO performance as the dependent variable. First the Pearson correlation is r (100) = 0.40 (Sig. 2-sided 0.000) which means that we have a positive correlation between them.

At the model summary in Table 13, the R^2 is 0.165 meaning that 16.5% of the total variance of SEO performance is due to website quality and error estimation is 5.48.

Table 13. Model summary between SEO performance and website quality.

Model	R	R Square	Adjusted R Square	Std. Error of the Estimate	Change Statistics	
					R Square Change	F Change
1	.406 [a]	.165	.156	5.48265941	.165	19.360

a. Predictors: (Constant), Shanghai rating.

The ANOVA examines the null hypothesis that there is no linear relationship between SEO performance and website quality. The result F (1.98) = 19.36 p = 0.000 < 0.05 is statistically significant so we reject the null hypothesis.

In the Table 14 it is depicted that the value of B is 0.22 (0.0000 < 0.01) which means that variable website quality rating can be a predictor of variable SEO performance. Also, there is a positive correlation between the two variables.

Table 14. Model coefficients simple regression analysis of website quality and SEO performance.

Model		Unstandardized Coefficients		Standardized Coefficients	t	Sig.
		B	Std. Error	Beta		
1	(Constant)	66.213	3.749		17.661	.000
	website quality rating	.227	.052	.406	4.400	.000

These results are visualized in Figure 2.

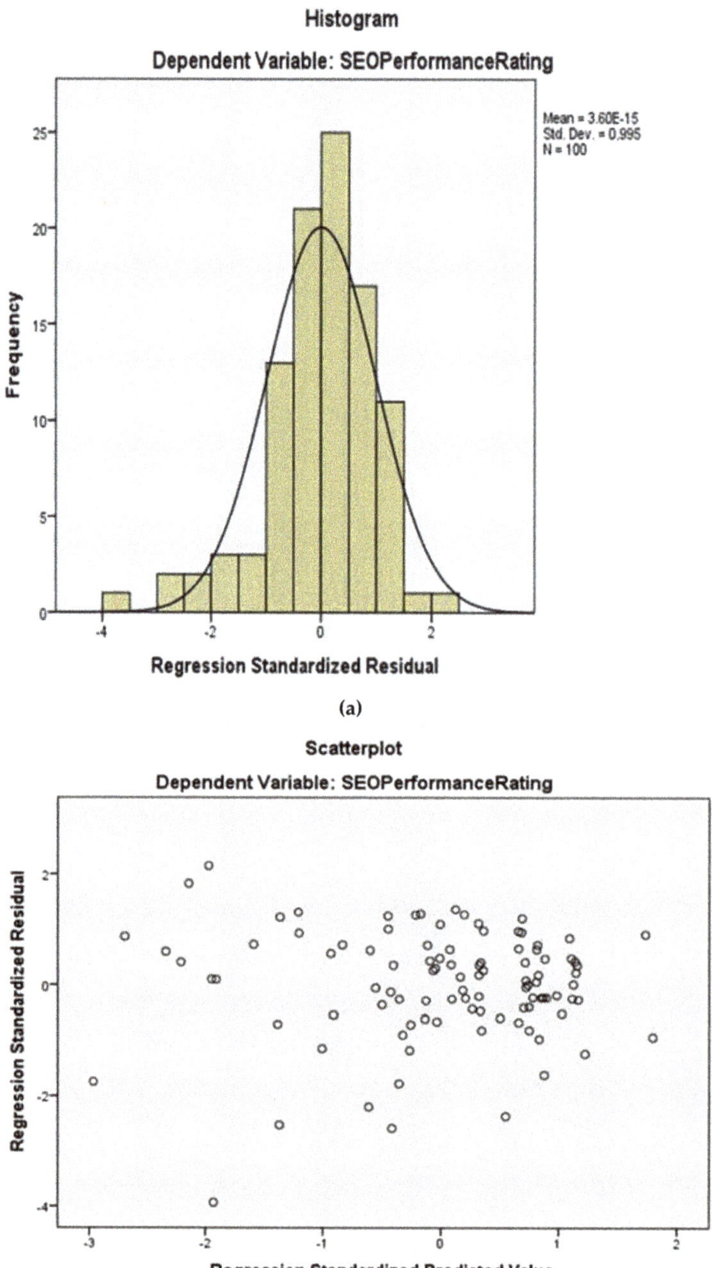

Figure 2. (a) The histogram of the standardized residuals shows a normal distribution. (b) The scatterplot shows that this is an acceptable predictive model (the prediction is on the x-axis, and the accuracy is on the y-axis - Residual = Observed − Predicted).

5. Discussion

Webometrics [17,68] is probably the most well-known approach in producing a ranking of universities taking into account their web presence. In a similar approach the same authors [69] attempted to evaluate institutional repositories as well as other open repositories that contain scientific literature (e.g., Arxiv.org and Hal CNRS, Citeseer). The indicators they employed for the evaluation of repositories were mostly targeting the activity and visibility of the website, where the activity accounts the number of pages, pdf files and items in Google Scholar, and the visibility the incoming links received by the repository. The earlier framework for webometrics from Björneborn and Ingwersen [70] used a similar methodology that was based on in and out links only.

Recent works by Hasan [71], Kaur et al. [72], and Nagpal et al. [73] examine online presence and content for academic institutions. In [71] the author successfully lists various criteria that are used by university ranking systems. It is noticeable that ranking systems who try to use an institution's web presence as a metric mainly focus on the volume of that presence. This volume is measured by activity, in and out links, high visibility in search engine results and other similar metrics that aim to determine the interconnectivity between the institutions and the general audience.

This study's approach is different. Firstly, the focus shifts from popularity metrics, such as activity and visibility, to quality metrics such as usability, accessibility and performance. Secondly, instead of trying to evaluate the institutions based on different aspects of their web presence it focused on evaluating the main website of each institution itself. This is less complicated to achieve since there are relatively objective criteria related with evaluating a website, both for its quality or for other aspects like its SEO performance. The study then proceeded to analyze whether these specific aspects of an Institution's online presence are related to its academic excellence ranking.

The present work leads relatively safely to the conclusion that academic excellence, as generally acknowledged in the Shanghai list, correlates with website quality, but not with SEO performance.

The overall web ranking confirms the main conclusion: this ranking has emerged as an average rating and its correlation with the Shanghai list is positively influenced by website quality and negatively by SEO performance. As the impact of website quality is stronger, the final result shows a positive, though not as strong, correlation.

Generally speaking, there is a positive correlation between academic excellence and the web presence of an institution. This observation is justified by the important role the Internet is currently serving as the basic medium for the development and implementation of communication policy at all levels, especially in organizations such as universities. With the further penetration of higher education into both societies and economies, with the increasing number of people attending university education and training programs, the resonance, the importance and, ultimately, the impact of website quality are evident. Universities are growing, their turnover is increasing, their funding and accountability needs are also increasing and, therefore, what they need now is not only to present their structures online, but engage daily, through their news and the digital storytelling of their activities, large sections of both the general and specific population. To this end, only the quality of their online presence can meet the aforementioned needs. That is why academic excellence and the online presence of an Institution are ultimately correlated.

Although the study of [71] examines only a limited set of sites shows that highly ranked universities on webometric sites have the lowest number of usability problems per investigated pages. The present study reinforces the notion that sites of institutions that rank high tend to display better metrics in aspects such as accessibility and usability.

The high correlation between the website quality ranking and SEO performance ranking—in the case of Institutions included in the Shanghai list—confirms that website quality is a factor that can give us a predictive or causal relationship with SEO performance.

From the aforementioned remarks, it appears as a reasonable interpretation of the results, that universities place particular emphasis on issues concerning website quality, while the emphasis on SEO matters does not appear to be of equal importance. This is explained by the fact that the tool we have presented includes evaluations of issues related to international standards (which the academic community endorses), open technologies, socially sensitive issues such as the level of accessibility and inclusion, as well as technological issues highlighting the technological infrastructure of each institution and its integration into the environment of new technologies.

However, the interpretation of the low degree of correlation with SEO performance probably has to do with the fact that SEO has strong marketing features and is highly market oriented. Furthermore, it is constantly used by commercial companies. It is conceivable that universities with a long-lasting presence and a well-established position in the international academic environment may not regard SEO as an important factor of their online presence or treat it more conventionally or as a tool only for commodities and not for 'goods' in the broader sense.

Nevertheless, in recent years the expansion of higher education and the need to reach wider audiences has made SEO an important factor in the online presence of universities. It might be the case that SEO is a feature that has not yet been fully exploited and can, with appropriate adaptations, make a difference for higher education institutions.

The present work also highlights the need for future work on two axes:

The first one is tool improvement. This may entail minor adjustments to the website quality tool, thorough examination of the SEO tool (because it had little scope for differential ranking) and reassessment of the composition through a simple average of the extent of the overall web presence or through normalization depending on the number of IT/press/Web development staff.

The second axis is comparison of website quality results with different aspects of academic excellence. Firstly, some of the different official rankings mentioned above. The Shanghai ranking is based purely in research metrics, but other rankings also take into consideration different aspects of academic excellence. For example, the teaching qualities of an institution (number of professors per student, teaching hours etc.) or aspects of its global presence. Additionally, there would be value in seeing how website quality correlates to the research efficiency of an institution. Research efficiency is a way of drawing comparisons between the input (as measured by grants etc.) and research output (articles etc.) and is an increasingly popular measurement of the excellence of an academic institution [74]. Finally, there may be conclusions to be drawn by monitoring more localized rankings and comparing the correlation between WQ and academic excellence in different parts of the world.

6. Conclusions

The present study documented the most important international academic university rankings and selected the Academic Ranking of World Universities (ARWU—also known as Shanghai list) as a base of reference, since it has been proven quite reliable for many years and consists of fully measurable and verifiable ratings.

Consequently, the characteristics of a) website quality and b) SEO performance were identified and the universities of the Shanghai list were ranked into a website quality ranking and a SEO performance ranking. In addition, the average of the two ratings was defined as the overall web ranking.

From the correlation of the traditional academic ranking of Shanghai and the website quality ranking and SEO performance ranking emerged some interesting observations. The positive correlation between academic excellence in general and the online presence of an institution (and especially its website quality) has now become apparent. On the other hand, no correlation was found with SEO performance. This means that the academic performance of an institution is correlated with the quality of a university's online presence; yet, obviously, this does not indicate a causal relation. Additionally, there seems to be significant room for improvement in regard to better use of SEO, always in accordance with academic criteria. This may present an advantage to the general presence and status of these institutions.

Even though further research is required to solidify the results of the present study, there is a clear indication that the web presence of academic institutions, as part of their wider communication policy, bears great significance not only for the institutions themselves, as it gives prominence to the institutions' global role, but also for the international community as it facilitates access for each and every individual to content related to higher education.

Supplementary Materials: Giannakoulopoulos, A; Konstantinou, N; Koutsompolis, D; Pergantis, M; Varlamis, I. Data set of the article: academic excellence, website quality, SEO performance: Is there a Correlation? (Version 1) (Data set). Zenodo. Available online: https://doi.org/10.5281/zenodo.3484673 (Accessed on 30 September 2019).

Author Contributions: Conceptualization, A.G.; formal analysis, N.K.; investigation, D.K. and M.P.; methodology, A.G. and M.P.; resources, I.V.; writing—original draft, N.K., D.K. and M.P.; writing—review and editing, A.G., M.P. and I.V.

Funding: This research received no external funding.

Conflicts of Interest: The authors declare no conflict of interest.

References

1. Mmantsetsa, M.; Wells, P.J.; Silvia, F. University Rankings: The Many Sides of the Debate. *Manag. Sustain. Dev.* **2014**, *6*, 39–42. [CrossRef]

2. Baka, A.B.A.; Leyni, N. Webometric study of world class universities websites. In *Qualitative and Quantitative Methods in Libraries*; Special Issue: Bibliometrics; 2017; pp. 105–115. Available online: http://qqml-journal.net/index.php/qqml/article/view/367 (accessed on 18 November 2019).

3. Bell, M.; Bush, D.; Nicholson, P.; O'Brien, D.; Tran, T. Universities online: A survey of online education and services in Australia. *Occas. Pap. Ser.* **2002**, *02-A*, 1–45.

4. Will, E.M.; Callison, C. Web presence of universities: Is higher education sending the right message online? *Public Relat. Rev.* **2006**, *32*, 180–183. [CrossRef]

5. McAllister-Spooner, S.M. Whose site is it anyway? Expectations of college Web sites. *Public Relat. J.* **2010**, *4*, 22–37.

6. Ford, W.G. Evaluating the Effectiveness of College Web Sites for Prospective Students. *J. Coll. Admiss.* **2011**, *212*, 26–31.

7. Abrahamson, T. Life and death on the Internet: To web or not to web is no longer a question. *J. Coll. Admiss.* **2000**, *6*, 168.

8. Kaur, S.; Kaur, K.; Kaur, P. An empirical performance evaluation of universities website. *Int. J. Comput. Appl.* **2016**, *146*, 53–62. [CrossRef]

9. Khwaldeh, S.; Al-Hadid, I.; Masa'deh, R.E.; Alrowwad, A.A. The association between e-services web portals information quality and ICT competence in the Jordanian universities. *Asian Soc. Sci.* **2017**, *13*, 156–169. [CrossRef]

10. Conway, M.; Dorner, D. An evaluation of New Zealand political party websites. *Inf. Res.* **2004**, *9*, 9-4.

11. Acosta-Vargas, P.; Luján-Mora, S.; Salvador-Ullauri, L. Evaluation of the web accessibility of higher-education websites. In Proceedings of the 15th International Conference on Information Technology Based Higher Education and Training (ITHET), Istanbul, Turkey, 8–10 September 2016; pp. 1–6.

12. Alahmadi, T.; Drew, S. An evaluation of the accessibility of top-ranking university websites: Accessibility rates from 2005 to 2015. In Proceedings of the DEANZ Biennial Conference, Waikato, New Zealand, 17–20 April 2016; pp. 224–233.

13. Beel, J.; Gipp, B.; Wilde, E. Academic Search Engine Optimization (aseo) Optimizing Scholarly Literature for Google Scholar & Co. *J. Sch. Publ.* **2009**, *41*, 176–190.

14. Rovira, C.; Codina, L.; Guerrero-Solé, F.; Lopezosa, C. Ranking by relevance and citation counts, a comparative study: Google Scholar, Microsoft Academic, WoS and Scopus. *Future Internet* **2019**, *11*, 202. [CrossRef]

15. Ale Ebrahim, N. Optimize Your Article for Search Engine. *Univ. Malaya Res. Bull.* **2014**, *2.1*, 38–39. [CrossRef]

16. Dimopoulos, M.A.; Mpourazelis, K.; Mpourletidis, K.; Koutsompolis, D.; International University Rankings. *Methodology – Indicators and the Position of NKUA 2017–2018*; National and Kapodistrian University of Athnes: Athens, Greece, 2018.

17. Aguillo, I.; Bar-Ilan, J.; Levene, M.; Ortega, J. Comparing university rankings. *Scientometrics* **2010**, *85*, 243–256. [CrossRef]

18. Academic Ranking of World Universities (ARWU). Available online: http://www.shanghairanking.com/arwu2019.html (accessed on 20 September 2019).

19. Liu, N.; Cheng, Y. Academic Ranking of World Universities—Methodologies and Problems. *High. Educ. Eur.* **2005**, *30*, 127–136. [CrossRef]

20. Webometrics. Available online: http://www.webometrics.info (accessed on 20 September 2019).

21. Times Higher Education Rankings. Available online: https://www.timeshighereducation.com/content/world-university-rankings (accessed on 20 September 2019).

22. QS World University Rankings. Available online: https://www.qs.com/rankings/ (accessed on 20 September 2019).

23. Center for World University Rankings (CWUR). Available online: https://cwur.org/about.php (accessed on 20 September 2019).

24. Bastedo, M.N.; Bowman, N.A. US News & World Report college rankings: Modeling institutional effects on organizational reputation. *Am. J. Educ.* **2009**, *116*, 163–183.

25. Ehrenberg, R.G. Reaching for the brass ring: The US News & World Report rankings and competition. *Rev. High. Educ.* **2003**, *26*, 145–162.

26. Huang, M.H. A comparison of three major academic rankings for world universities: From a research evaluation perspective. *J. Libr. Inf. Stud.* **2011**, *9*, 1–25.

27. Academic Ranking of World Universities: Methodology. Available online: http://www.shanghairanking.com/ARWU-Methodology-2019.html (accessed on 20 September 2019).

28. ShanghaiRanking Academic Excellence Survey 2019 Methodology. Available online: http://www.shanghairanking.com/subject-survey/survey-methodology-2019.html (accessed on 20 September 2019).

29. Why Validate? Available online: https://validator.w3.org/docs/why.html (accessed on 17 September 2019).

30. W3C's Markup Validation Service. Available online: https://validator.w3.org/ (accessed on 17 September 2019).

31. World Wide Web Consortium. Available online: https://www.w3.org/Consortium/ (accessed on 17 September 2019).

32. About The W3C Markup Validation Service. Available online: https://validator.w3.org/about.html (accessed on 17 September 2019).

33. About the CSS Validator. Available online: https://jigsaw.w3.org/css-validator/about.html (accessed on 17 September 2019).

34. CSS Validation Service. Available online: https://jigsaw.w3.org/css-validator/ (accessed on 17 September 2019).

35. Conformance: Requirements and Recommendations. Available online: https://www.w3.org/TR/CSS21/conform.html#valid-style-sheet (accessed on 17 September 2019).

36. Enge, E. Where is the Mobile vs. Desktop Story Going? Perficient Digital. 2019. Available online: https://www.perficientdigital.com/insights/our-research/mobile-vs-desktop-usage-study (accessed on 17 September 2019).

37. Mobile-Friendly Test. Available online: https://search.google.com/test/mobile-friendly (accessed on 17 September 2019).

38. Support Google, Mobile-Friendly Test Tool. Available online: https://support.google.com/webmasters/answer/6352293 (accessed on 15 September 2019).

39. Developers Google, Lighthouse Scoring Guide. Available online: https://developers.google.com/web/tools/lighthouse/v3/scoring (accessed on 15 September 2019).

40. Developers Google, Lighthouse. Available online: https://developers.google.com/web/tools/lighthouse (accessed on 16 September 2019).

41. Introduction to Web Accessibility. Available online: https://www.w3.org/WAI/fundamentals/accessibility-intro/ (accessed on 15 September 2019).

42. Web Content Accessibility Guidelines (WCAG) Overview. Available online: https://www.w3.org/WAI/standards-guidelines/wcag/#intro (accessed on 15 September 2019).

43. AC Checker. Available online: https://achecker.ca/checker/index.php (accessed on 15 September 2019).

44. Wave. Available online: https://wave.webaim.org/ (accessed on 16 September 2019).

45. WebAIM. Available online: https://webaim.org/about/ (accessed on 16 September 2019).

46. Web Accessibility Evaluation Tools List. Available online: https://www.w3.org/WAI/ER/tools/ (accessed on 16 September 2019).

47. Wagner, J. Why Performance Matters. WebFundamentals. Available online: https://developers.google.com/web/fundamentals/performance/why-performance-matters (accessed on 16 September 2019).

48. WebPagetest. Available online: https://www.webpagetest.org/ (accessed on 18 September 2019).

49. About WebPagetest.org. Available online: https://www.webpagetest.org/about (accessed on 18 September 2019).

50. WebPagetest Documentation. Available online: https://sites.google.com/a/webpagetest.org/docs/using-webpagetest/metrics (accessed on 18 September 2019).
51. Apache. Available online: https://httpd.apache.org/docs/2.4/mod/core.html (accessed on 18 September 2019).
52. Compression in HTTP. Available online: https://developer.mozilla.org/en-US/docs/Web/HTTP/Compression (accessed on 18 September 2019).
53. Grigorik, I. Image Optimization. WebFundamentals. Available online: https://developers.google.com/web/fundamentals/performance/optimizing-content-efficiency/image-optimization (accessed on 18 September 2019).
54. Filestack. Why Progressive Images Matter. Available online: https://blog.filestack.com/thoughts-and-knowledge/progressive-images-matter/ (accessed on 18 September 2019).
55. WebPagetest Documentation. Available online: https://sites.google.com/a/webpagetest.org/docs/using-webpagetest/quick-start-quide (accessed on 18 September 2019).
56. About PageSpeed Insights. Available online: https://developers.google.com/speed/docs/insights/v5/about (accessed on 18 September 2019).
57. GitHub. Available online: https://github.com/GoogleChrome/lighthouse/blob/master/docs/throttling.md (accessed on 18 September 2019).
58. Giomelakis, D.; Veglis, A. Search Engine Optimization. In *Encyclopedia of Information Science and Technology*; IGI Global: Hershey, PA, USA, 2018; Available online: https://www.academia.edu/33637489/Search_Engine_Optimization (accessed on 18 September 2019).
59. Neil Patel. Available online: https://neilpatel.com/ (accessed on 18 September 2019).
60. Neil Patel Seo Analyzer. Available online: https://neilpatel.com/seo-analyzer/ (accessed on 18 September 2019).
61. SEO Site Checkup. Available online: https://seositecheckup.com/ (accessed on 18 September 2019).
62. SEO Site Checkup—About Us. Available online: https://seositecheckup.com/about (accessed on 18 September 2019).
63. WooRank. Available online: https://www.woorank.com/ (accessed on 18 September 2019).
64. WooRank Index: Websites Listed by Countries and Technology. Available online: https://index.woorank.com/en/reviews (accessed on 18 September 2019).
65. Roussos, P.L.; Tsaousis, G. *Statistics in Behavioural Sciences Using SPSS*; TOPOS: Athens, Greece, 2011.
66. Seltman, H.J. *Experimental Design and Analysis*; Carnegie Mellon University: Pittsburgh, PA, USA, 2012; Available online: http://www.stat.cmu.edu/~{}hseltman/309/Book/Book.pdf (accessed on 29 September 2019).
67. Lane, D.M. Online statistics education: An interactive multimedia course of study. *The Independent*, 2015.
68. Aguillo, I.F.; Ortega, J.L.; Fernández, M. Webometric ranking of world universities: Introduction, methodology, and future developments. *High. Educ. Eur.* **2008**, *33*, 233–244. [CrossRef]
69. Aguillo, I.; Ortega, J.; Fernández, M.; Utrilla, A. Indicators for a webometric ranking of open access repositories. *Scientometrics* **2010**, *82*, 477–486. [CrossRef]
70. Björneborn, L.; Ingwersen, P. Toward a basic framework for webometrics. *J. Am. Soc. Inf. Sci. Technol.* **2004**, *55*, 1216–1227. [CrossRef]
71. Hasan, L. Using university ranking systems to predict usability of university websites. *JISTEM-J. Inf. Syst. Technol. Manag.* **2013**, *10*, 235–250. [CrossRef]
72. Kaur, S.; Kaur, K.; Kaur, P. Analysis of website usability evaluation methods. In Proceedings of the 2016 3rd International Conference on Computing for Sustainable Global Development (INDIACom), New Delhi, India, 16–18 March 2016; BVICAM: New Delhi, India; pp. 1043–1046.
73. Nagpal, R.; Mehrotra, D.; Bhatia, P.K.; Sharma, A. Rank university websites using fuzzy AHP and fuzzy TOPSIS approach on usability. *Int. J. Inf. Eng. Electron. Bus.* **2015**, *7*, 29. [CrossRef]
74. Wohlrabe, K.; Bornmann, L.; de Moya Anegon, F. How efficiently produce elite US universities highly cited papers? *Publications* **2019**, *7*, 4. [CrossRef]

 *future internet*

Article

Ranking by Relevance and Citation Counts, a Comparative Study: Google Scholar, Microsoft Academic, WoS and Scopus

Cristòfol Rovira *, **Lluís Codina**, **Frederic Guerrero-Solé and Carlos Lopezosa**

Department of Communication, Universitat Pompeu Fabra, 08002 Barcelona, Spain; lluis.codina@upf.edu (L.C.); frederic.guerrero@upf.edu (F.G.-S.); carlos.lopezosa@upf.edu (C.L.)
* Correspondence: cristofol.rovira@upf.edu; Tel.: +34-667295308

Received: 5 August 2019; Accepted: 14 September 2019; Published: 19 September 2019

Abstract: Search engine optimization (SEO) constitutes the set of methods designed to increase the visibility of, and the number of visits to, a web page by means of its ranking on the search engine results pages. Recently, SEO has also been applied to academic databases and search engines, in a trend that is in constant growth. This new approach, known as academic SEO (ASEO), has generated a field of study with considerable future growth potential due to the impact of open science. The study reported here forms part of this new field of analysis. The ranking of results is a key aspect in any information system since it determines the way in which these results are presented to the user. The aim of this study is to analyze and compare the relevance ranking algorithms employed by various academic platforms to identify the importance of citations received in their algorithms. Specifically, we analyze two search engines and two bibliographic databases: Google Scholar and Microsoft Academic, on the one hand, and Web of Science and Scopus, on the other. A reverse engineering methodology is employed based on the statistical analysis of Spearman's correlation coefficients. The results indicate that the ranking algorithms used by Google Scholar and Microsoft are the two that are most heavily influenced by citations received. Indeed, citation counts are clearly the main SEO factor in these academic search engines. An unexpected finding is that, at certain points in time, Web of Science (WoS) used citations received as a key ranking factor, despite the fact that WoS support documents claim this factor does not intervene.

Keywords: ASEO; SEO; reverse engineering; citations; google scholar; microsoft academic; web of science; WoS; scopus; indicators; algorithms; relevance ranking; citation databases; academic search engines

---

1. Introduction

The ranking of search results is one of the main challenges faced by the field of information retrieval [1,2]. Search results are sorted so that the results best able to solve the user's need for information are ranked at the top of the page [3]. The challenges faced though are far from straightforward given that a successful ranking by relevance depends on the correct analysis and weighting of a document's properties, as well as the analysis of the need for that information and the key words used [1,2,4].

Relevance ranking has been successfully employed in a number of areas, including web page search engines, academic search engines, academic author rankings and the ranking of opinion leaders on social platforms [5]. Many algorithms have been proposed to automate this relevance and some of them have been successfully implemented. In so doing, different criteria are applied depending on the specific characteristics of the elements to be ordered. PageRank [6] and Hyperlink-Induced Topic Search (HITS) [7] are the best know algorithms for ranking web pages. Variants of these algorithms have also been used to rank influencers in social media, and include, for example, IP-Influence [8], TunkRank [9], TwitterRank [10] and TURank [11]. To search for academic documents, various algorithms have been proposed and used, both for the documents themselves and for their authors. These include Authority-Based Ranking [12], PopRank [13], Browsing-Based Model [14] and CiteRank [15]. All of them use the number of citations received by the articles as a search ranking factor in combination with other elements, such as publication date, the author's reputation and the network of relationships between documents, authors and affiliated institutions.

Many information retrieval systems (search engines, bibliographic databases and citation databases, etc.) use relevance ranking in conjunction with other types of sorting, including chronological, alphabetical by author, number of queries and number of citations. In search engines like Google, relevance ranking is the predominant approach and is calculated by considering more than 200 factors [16,17]. Unfortunately, Google does not release precise details about these factors, it only publishes fairly sketchy, general information. For example, the company says that inbound links and content quality are important [18,19]. Google justifies this lack of transparency in order to fight search engine spam [20] and to prevent low quality documents from being ranked at the top of the results by falsifying their characteristics.

Search engine optimization (SEO) is the discipline responsible for optimizing websites and their content to ensure they are ranked at the top of the search engine results pages (SERPs), in accordance with the relevance ranking algorithm [21]. In recent years, SEO has also been applied to academic search engines, such as Google Scholar and Microsoft Academic. This new application has received the name of "academic SEO" (or ASEO) [22–26]. ASEO helps authors and publishers to improve the visibility of their publications, thus increasing the chances that their work will be read and cited.

However, it should be stressed that the relevance ranking algorithm of academic search engines differs from that of standard search engines. The ranking factors employed by the respective search engine types are not the same and, therefore, many of those used by SEO are not applicable to ASEO while some are specific to ASEO (see Table 1).

SEO companies [27–29] routinely conduct reverse engineering research to measure the impact of the factors involved in Google's relevance ranking. Based on the characteristics of the pages that appear at the top of the SERPs, the factors with the greatest influence on the relevance ranking algorithm can be deduced. It is not a straightforward task since many factors have an influence and, moreover, the algorithm is subject to constant changes [30].

Studies that have applied a reverse engineering methodology to Google Scholar have shown that citation counts are one of the key factors in relevance ranking [31–34]. Microsoft Academic, on the other hand, has received less attention from the scientific community [35–38] and there are no specific studies of the quality of its relevance ranking.

Academic search engines, such as Google Scholar and Microsoft Academic, are an alternative to bibliographic commercial databases, such as Web of Science (WoS) and Scopus, for indexing scientific citations and they provide a free service of similar performance that competes with the business model developed by the classic services. Unlike search engines, bibliographic databases are fully transparent about how they calculate relevance, clearly informing users how their algorithm works on their help pages [39,40].

The primary aim of this study is to verify the importance attached to citations received in the relevance ranking algorithms of two academic search engines and two bibliographic databases. We analyze the two main academic search engines (i.e., Google Scholar and Microsoft Academic) and the two bibliographic databases of citations providing the most comprehensive coverage (WoS and Scopus) [41].

We address the following research questions: Is the number of citations received a key factor in Google Scholar relevance rankings? Do the Microsoft Academic, WoS and Scopus relevance algorithms operate in the same way as Google Scholar's? Do citations received have a similarly strong influence on all these systems? A similar approach to the one adopted here has been taken in previous studies of the factors involved in the ranking of scholarly literature [22,23,31–34].

The rest of this manuscript is organized as follows. First, we review previous studies of the systems that concern us here, above all those that focus on ranking algorithms. Next, we explain the research methodology and the statistical treatment performed. We then report, analyze and discuss the results obtained before concluding with a consideration of the repercussions of these results and possible new avenues of research.

2. Related Studies

Google Scholar, Microsoft Academic, WoS and Scopus have been analyzed previously in works that have adopted a variety of approaches, including, most significantly:

- Comparative analyses of the coverage and quality of the academic search engines and bibliographic databases [42–51]
- Studies of the impact of authors and the h-index [33,44,52–57]
- Studies of the utility of Google Scholar and Academic Search for bibliometric studies [20,49,55,58–61]

However, few studies [43,62] have focused their attention on information retrieval and the search efficiency of academic search engines, while even fewer papers [22,23,31–34] have examined the factors used in ranking algorithms.

The main conclusions to be drawn from existing studies of relevance ranking in the systems studied can be summarized as follows:

- The number of citations received is a very important factor in Google Scholar relevance rankings, so that documents with a high number of citations received tend to be ranked first [32–34].
- Documents with many citations received have more readers and more citations and, in this way, consolidate their top position [61].

Surprisingly, the relevance ranking factors of academic search engines and bibliographic databases have attracted little interest in the scientific community, especially if we consider that a better position in their rankings means enhanced possibilities of being found and, hence, of being read. Indeed, the initial items on a SERP have been shown to receive more attention from users than that received by items lower down the page [63].

Table 1. Search engine optimization (SEO) and academic search engine optimization (ASEO) factors. WoS, Web of Science.

Type	SEO/ASEO factor	Google Search	Google Scholar	Microsoft Academic	WoS	Scopus
On-page factors	Keywords in title	Yes [16–19,28–30]	Yes [31,32]	?	Yes [40]	Yes [41]
	Keywords in URL, h1 or first words	Yes [16–19,28–30]	?	?	No [40]	No [41]
	Keyword frequency	No [16–19]	?	?	Yes [40]	Yes [41]
	Technical factors: design, speed, etc.	Yes [16–19,28–30]	?	?	No [40]	No [41]
Off-page factors	Backlinks	Yes [16–19,28–30]	?	?	No [40]	No [41]
	Received citations	?	Yes [16,31–34]	Yes [35–38,64]	No [40]	No [41]
	Author reputation	Yes [16–19,28–30]	Yes [16]	Yes [35–38,64]	No [40]	No [41]
	Reputation of the publication or domain	Yes [16–19,28–30]	Yes [16]	Yes [35–38,64]	No [40]	No [41]
	Signals from social networks	Yes (Indirect) [16–19]	?	?	No [40]	No [41]
	Traffic, Click Through Rate	Yes [16–19,28–30]	?	?	No [40]	No [41]
Artificial intelligence	RankBrain	Yes [17–19,28–30]	?	?	No [40]	No [41]

In the light of these previous reports, it can be concluded that the number of intervening factors in the academic search engines is likely to be fewer than those employed by Google and that, therefore, the algorithm is simpler (see Table 1).

3. Methodology

This study is concerned with analyzing the relevance ranking algorithms used by academic information retrieval systems. We are particularly interested in identifying the factors they employ, especially in the case of systems that do not explain how their ranking algorithm works. A reverse engineering methodology is applied to the two academic search engines (i.e., Google Scholar and Microsoft Academic) and to two bibliographic databases of citations (i.e., WoS and Scopus). These, in both cases, are the systems offering the most comprehensive coverage [41,65,66]. The specific objective is to identify whether the citations received by the documents are a determining factor in the ranking of search results.

Reverse engineering is a research method commonly used to study any type of device in order to identify how it works and what its components are. It is a relatively economical way to obtain information about the design of a device or the source code of a computer program based on the compiled files.

One of the fields in which reverse engineering is being applied most is precisely in that of detecting the factors included in Google's relevance ranking algorithm [28,29,67]. The little information provided by Google [16] is used as a starting point to analyze the characteristics of the pages ranked at the top of the search results to deduce what factors are included and what their respective weighting is. However, the ranking algorithms are complex [68], moreover, they are subject to frequent modifications and the results of reverse engineering are usually inconclusive. Recently, a reverse engineering methodology has also been applied to academic search engines [34].

To obtain an indication of the presence of a certain positioning factor, the ranking data are treated statistically by applying Spearman's correlation coefficient, selected here because the distribution is not normal according to Kolmogorov–Smirnov test results. Generally, a comparison is made of the ranking created by the researcher using the values of the factor under study with the search engine's natural ranking—for example, a ranking based on the frequency of the appearance of the keywords in the document and Google's native ranking. If a high coefficient is obtained, this means that this factor is contributing significantly to the ranking. However, in the case of Google, many factors intervene: more than 200, according to many sources [69,70]. Therefore, it is very difficult to detect high correlations indicative of the fact that a certain characteristic has an important weighting. Statistical studies generally consider a correlation between 0.4 and 0.7 to be moderate and a correlation above 0.7

to be high. In reverse engineering studies with Google, the correlation values between the positions of the pages and the quantitative values of the supposed positioning factors do not normally exceed 0.3 [68]. Although the correlations are low, with studies of this type, relatively clear indications of the factors intervening in the ranking can be obtained.

Google themselves provide even less data on how they rank by relevance in Google Scholar. Perhaps their most explicit statement is the following:

"Google Scholar aims to rank documents the way researchers do, weighing the full text of each document, where it was published, who it was written by as well as how often and how recently it has been cited in other scholarly literature." [71].

Previous research [64,72] has shown that Google Scholar applies far fewer ranking factors than is the case with Google's general search engine. This is a great advantage when applying reverse engineering since the statistical results are much clearer, with some correlations being as high as 0.9 [34].

Likewise, Microsoft Academic does not offer any specific details about its relevance ranking algorithm [73]. We do know, however, that it applies the Microsoft Academic Graph or MAG [74], an enormous knowledge database made up of interconnected entities and objects. A vector model is applied to identify the documents with the greatest impact using the PopRank algorithm [13,75]. However, Microsoft Academic does not indicate exactly what the "impact" is when this concept is applied to the sorting algorithm:

"In a nutshell, we use the dynamic eigencentrality measure of the heterogeneous MAG to determine the ranking of publications. The framework ensures that a publication will be ranked high if it impacts highly ranked publications, is authored by highly ranked scholars from prestigious institutions, or is published in a highly regarded venue in highly competitive fields. Mathematically speaking, the eigencentrality measure can be viewed as the likelihood that a publication will be mentioned as highly impactful when a survey is posed to the entire scholarly community" [76]

Unlike these two engines, the WoS and Scopus bibliographic databases provide detailed information about their relevance ranking factors [39,40]. In systems of this type, a vector model is applied [1] and relevance is calculated based on the frequency and position of the keywords of the searches in the documents; therefore, citations received are not a factor.

Another factor that facilitates the use of reverse engineering in the cases of Google Scholar, Microsoft Academic, WOS and Scopus is the information these systems provide regarding the exact number of citations received, a factor used to compute their rankings. Unlike the general Google search engine that does not give reliable information about the number of inbound links, in the four systems studied the number of citations received, and even the listing of all citing documents, is easily obtained. The relative simplicity of the algorithms and the accuracy of the citation counts mean reverse engineering is especially productive when applied to the study of the influence of citations in relevance ranking in academic search engines and bibliographic databases of citations.

For the study reported here, 25,000 searches were conducted in each system, a similar number to those typically conducted in reverse engineering studies [27–29] or other analyses of Google Scholar rankings [22,23,31–34]. The ranking provided by each tool was then compared with a second ranking created applying only the number of citations received. As the distributions were not normal according to Kolmogorov–Smirnov test results, Spearman's correlation coefficient was calculated. The hypothesis underpinning reverse engineering as applied to search engine results is that the higher the correlation coefficient, the more similar the two rankings are and, therefore, a greater weight can be attributed to the isolated factor used in the second ranking.

To avoid thematic biases, the keywords selected for use in the searches needed to be as neutral as possible. Thus, we chose the 25 most frequent keywords appearing in academic documents [77–79].

Searches were then conducted in Google Scholar and Microsoft Academic using these keywords which also enabled us to identify two-term searches based on the suggestions made by these two engines. Next, from among these suggestions we selected those with the greatest number of results. In this way, we obtained two sets of 25 keywords, the first formed by a single term and the second by two. The terms used can be consulted in Annex 1. It is critical that each search provides as many results as possible in order to ensure that the statistical treatment of the rankings of these results is robust. It is for this reason that we didn't use Long Tail Keywords.

To address our research question concerning the impact of citations received on ranking, we undertook searches with these keywords in each system collecting up to 1000 results each time. In the case of the academic search engines, searches were carried out with both one and two-term keywords. In the case of the bibliographic databases, searches were only carried out with two-term keywords since our forecasts were very clear in indicating that citations did not affect the results—indeed, the documentation for these systems also make this quite clear. However, as we see below, the results were not as expected.

The search engine data were obtained using the Publish or Perish tool [80,81] between 10 May 2019 and 30 May 2019 (see Appendices A and B). The data from the bibliographic databases were obtained by exportation from the systems themselves between these same dates.

In each of the systems studied, our rankings created using citation counts were compared with the native rankings of each system. To do this, the number of citations received was transformed to an ordinal scale, a procedure previously used in other studies [31,32]. According to reverse engineering methodology, if the two rankings correlate then they are similar and, therefore, it can be deduced that citations are an important factor in the relevance ranking algorithms. The ranking by citations received was correlated for each of the 25 searches (and their corresponding 1000 results) carried out in each system with the native ranking of these systems.

To obtain a global value for each system which integrates the 25 searches and their corresponding 25,000 data items, the median values of each of the 25 citation search rankings were used for each position in the native ranking. The native ranking positions of each system are shown on the x-axis, while the rankings according to citations received are shown on the y-axis. Each gray dot corresponds to one of the 25,000 data items from the 25 searches of 1000 results each conducted on each system. The blue dots are the 1000 median values that indicate the central tendency of the data. The more compact and the closer to the diagonal the medians are, the greater the correlation between the two rankings.

The software used in the analysis was R, version 3.4.0 [82] and SPSS, version 20.0.0. The confidence intervals were constructed via normal approximation by applying Fisher's transformation using the R psych package [83,84]. Fisher's transformation when applied to Spearman's correlation coefficient is asymptotically normal. Graphs were drawn with Google Sheets and Tableau.

4. Analysis of Results

Table 2 shows the results obtained when analyzing the four systems. It can be seen that in some cases different analyses were conducted on the same system. This reflects various circumstances impacting the study. For example, in the case of Microsoft Academic we did not perform a full-text search, rather it was limited to the bibliographic reference: that is, the title, keywords, name of publication, author and the abstract. Interestingly, in conducting the study we found that in more than 95% of the searches the keywords were present in the result titles. This gave rise to a problem when we compared the results with Google Scholar, since this search engine does perform full-text searches. For this reason, we undertook a second data collection in the case of Google Scholar, restricting searches to the title and, in this way, we are able to make a more accurate comparison of the results provided by Google Scholar and Microsoft Academic. The introduction of a second variant was a result of the number of search terms. The study was carried out using two sets of 25 keywords, the first made up of a single term and the second of two terms. Finally, in the case of WoS, two data collections were undertaken since it became apparent that the ranking criteria had changed. However, each of these variants allows us to make partial comparisons and to analyze specific aspects of the systems.

Table 2. Correlation coefficients for the academic search engines.

System	Number of Search Terms	Search Restrictions Included	Spearman's Coefficient	p
Google Scholar	1	unrestricted	0.968	<0.0001
Google Scholar	2	unrestricted	0.721	<0.0001
Google Scholar	1	title	0.990	<0.0001
Google Scholar	2	title	0.994	<0.0001
Microsoft Academic	1	title, abstract, keywords	0.907	<0.0001
Microsoft Academic	2	title, abstract, keywords	0.937	<0.0001
Scopus	2	title, abstract, keywords	−0.107	<0.001
WoS-version 1	2	title, abstract, keywords	−0.075	<0.05
Wos-version 2	2	title, abstract, keywords	0.907	<0.0001

The correlations for the two academic search engines were, in all cases, higher than 0.7 and in some cases reached 0.99 (Table 2). These results indicate that citation counts are likely to constitute a key ranking factor with considerable weight in the sorting algorithm. The other factors can cause certain results to climb or fall down the ranking, but the main factor would appear to be citations received. In the case of the bibliographic databases, the correlations were close to zero and, therefore, in such systems, we have no evidence that citations intervene. However, in the case of WoS, over a period of several days, we detected it to be using different sorting criteria and its results were in fact ranked using the number of citations received as its primary factor. This result is surprising since it does not correspond to the criteria the database claims to use in its corresponding documentation. We attribute these variations to tests being performed to observe user reactions, and on the basis of these results, decisions can presumably be taken to modify the ranking algorithm, but this is no more than an inference.

We find virtually no differences between searches conducted with either one or two terms in the academic search engines (see Figures 1–6). In principle, there is no reason as to why there should be any differences as the same ranking algorithm was applied in all cases. However, we did find some differences in the case of Google Scholar when searches were performed with or without restriction to the title, the same search providing a different ranking. When the terms were in all the titles, correlation coefficients of 0.9 were obtained (see Figures 3–6). When this was not the case, the correlation coefficient was still very high, but fell to 0.7 in the searches with two words (see Figure 2). This difference of almost 20 points is a clear indication that the inclusion of keywords in the title is also an important positioning factor. These differences are even more evident if we analyze the correlation coefficients of each search. Figure 7 shows how in all cases the correlation coefficients of the searches performed with this restriction are greater than when this restriction is not included. When restricting searches to the title, this factor is nullified since all the results have it. In such circumstances, the correlation is almost 1 since practically the only factor in operation is that of citations received. Therefore, indirectly, we can verify that the inclusion of the search terms in the document title forms part of the sorting algorithm, which we already knew, but for which we are now able to provide quantitative evidence showing that the weight of this factor is lower than that of citations since there is little difference between the two correlations. Unfortunately, this same analysis cannot be conducted in the case of Microsoft Academic because it does not permit full-text searches.

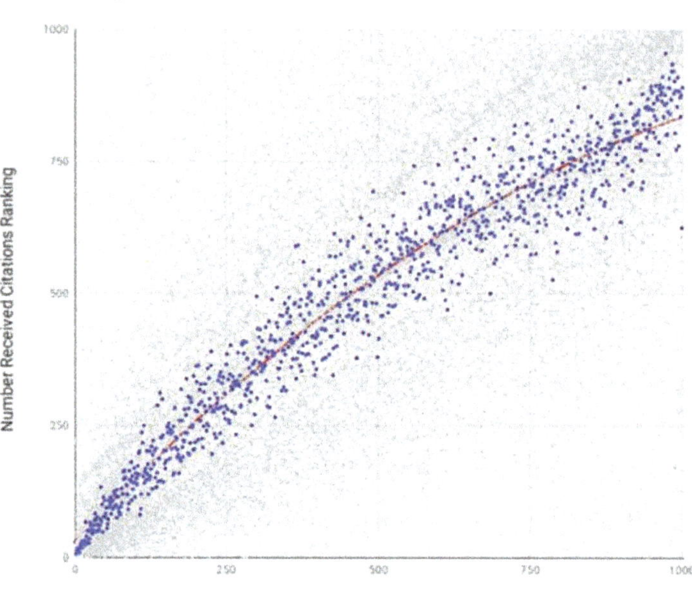

Figure 1. Google Scholar Searches with One Word and No Title Restriction (rho 0.968, median in blue, the rest of data in gray).

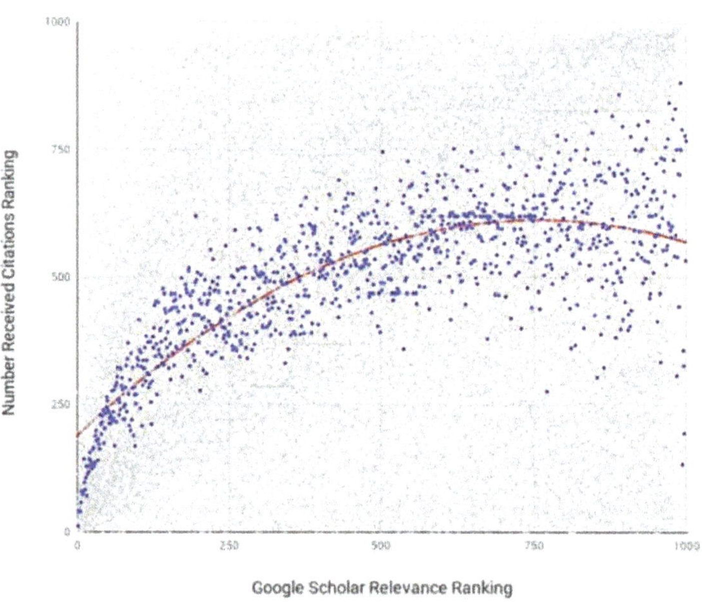

Figure 2. Google Scholar Searches with Two Words and No Title Restriction (rho 0.721, median in blue, the rest of data in gray).

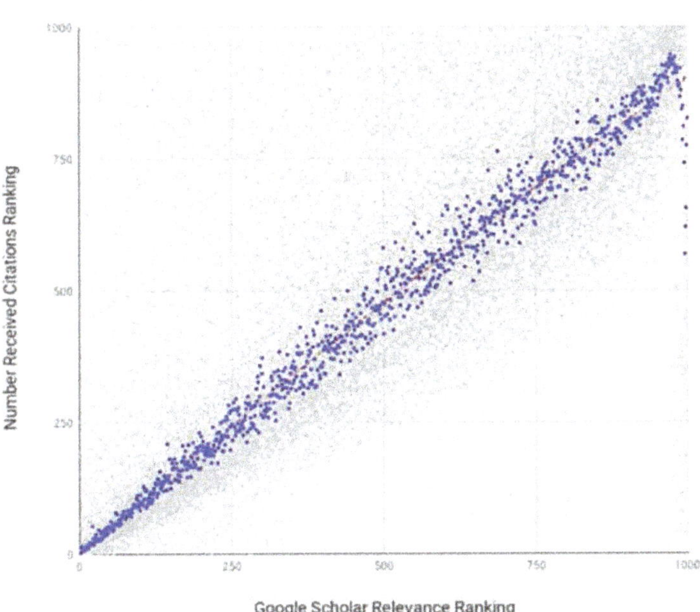

Figure 3. Google Scholar Searches with One Word and Title Restriction (rho 0.990, median in blue, the rest of data in gray).

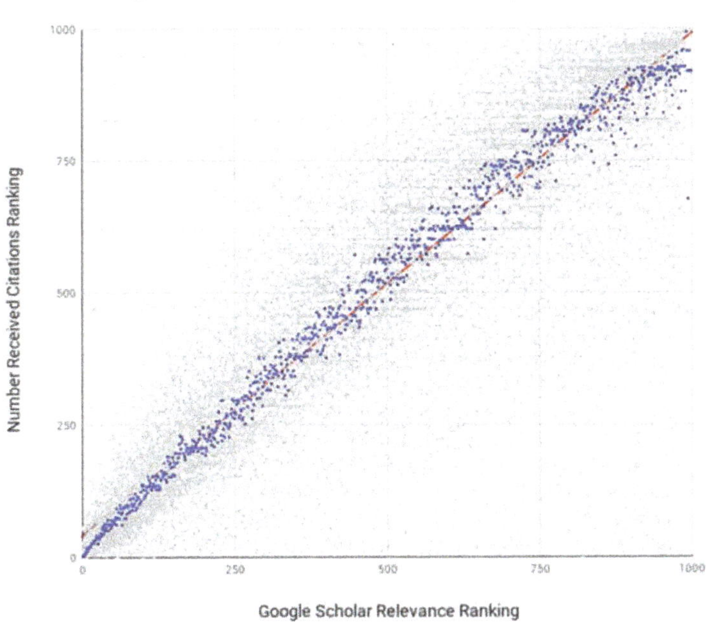

Figure 4. Google Scholar Searches with Two Words and Title Restriction (rho 0.994, median in blue, the rest of data in gray).

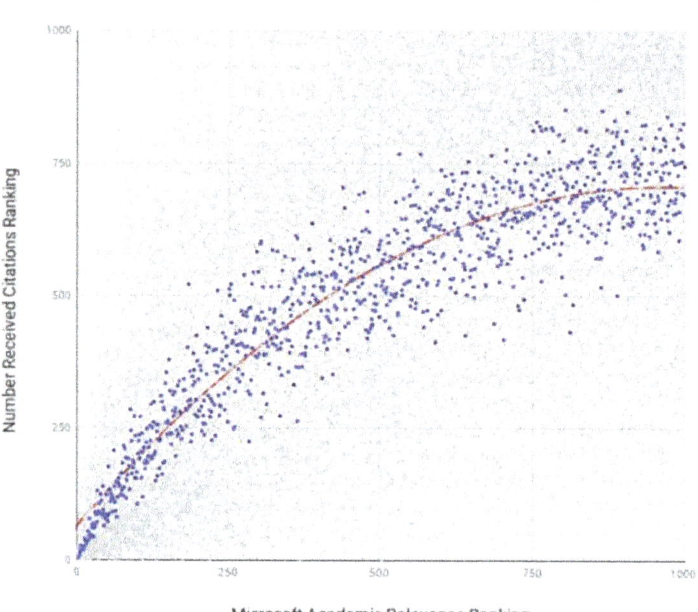

Figure 5. Microsoft Academic Searches with One Word and Title Restriction (rho 0.907, median in blue, the rest of data in gray).

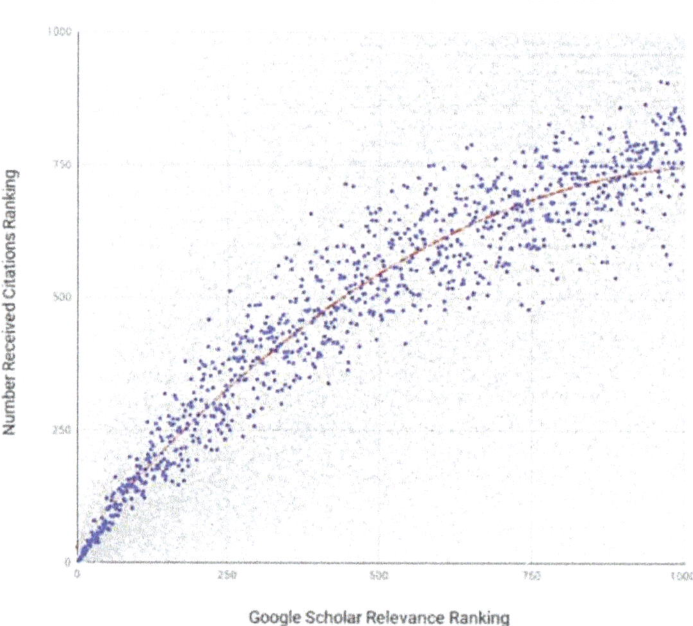

Figure 6. Microsoft Academic Searches with Two Words and Title Restriction (rho 0.937, median in blue, the rest of data in gray).

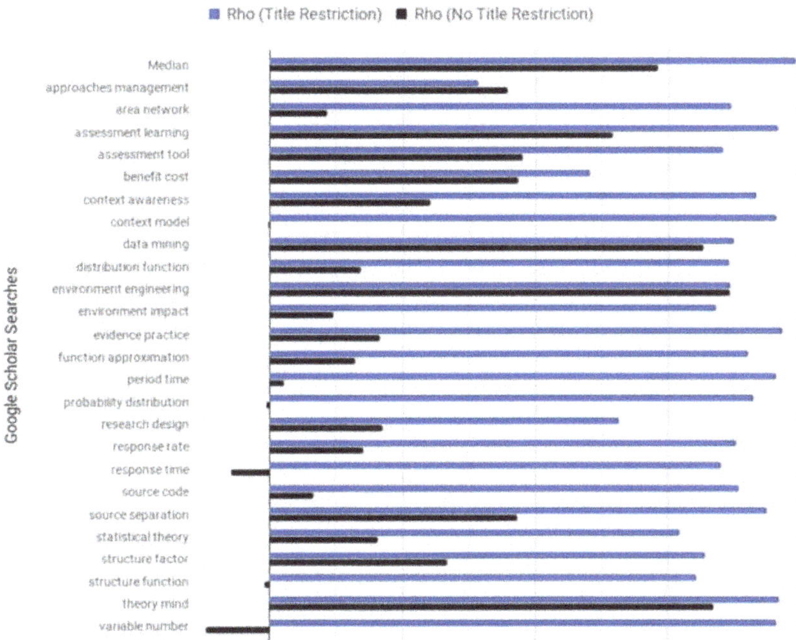

Figure 7. Google Scholar: Title Restriction vs No Title Restriction with Two Words.

Likewise, we detected no differences between Google Scholar and Microsoft Academic, above all when comparing the results of searches restricted to the title (Figures 3 and 6). In all four cases, we obtained correlation coefficients of 0.9. Therefore, it seems that both academic search engines apply a very similar weight to the factor of citations received.

Finally, it is worth noting that the two bibliographic databases (see Figures 8 and 9) do not employ citations received as a positioning factor, as stated in their documentation. Therefore, it is perfectly logical that their corresponding correlation coefficients are almost zero. However, somewhat surprisingly, in the case of WoS, we found that between 20 May 2019 and 25 May 2019 the ranking was different and we obtained a correlation coefficient similar to that of the academic search engines, i.e., 0.9 (see Figures 10–12) and that, therefore, a different algorithm was being applied with the significant inclusion of citation counts. It is common that before introducing changes in the design of websites, tests are made with real users. Figures 10 and 11 illustrate screenshots of the same search but employing two different relevance rankings. This can be done by randomly publishing different prototypes to gather information on user behavior in order to determine which prototype achieves greatest acceptance. As discussed above, it would seem that WoS was implementing such a procedure and was carrying out tests aimed at modifying its relevance ranking using an algorithm similar to that of academic search engines, although we should insist that this is only an inference.

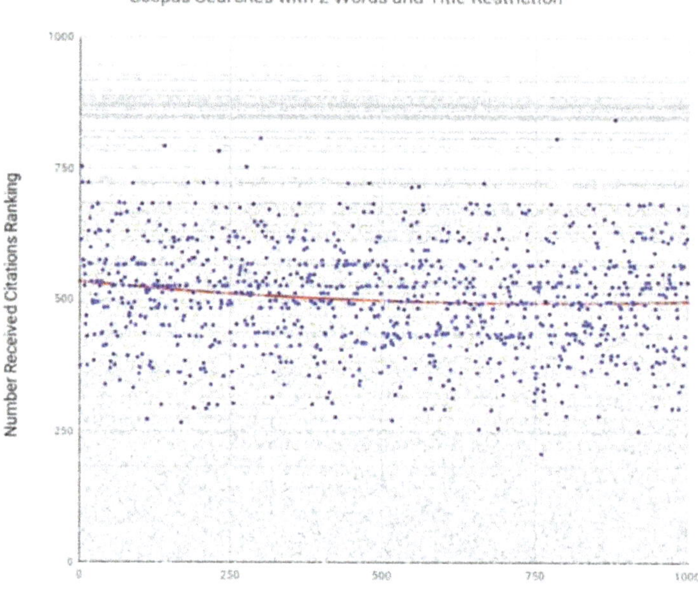

Figure 8. Scopus Searches with Two Words and Title Restriction (rho −0.10, median in blue, the rest of data in gray).

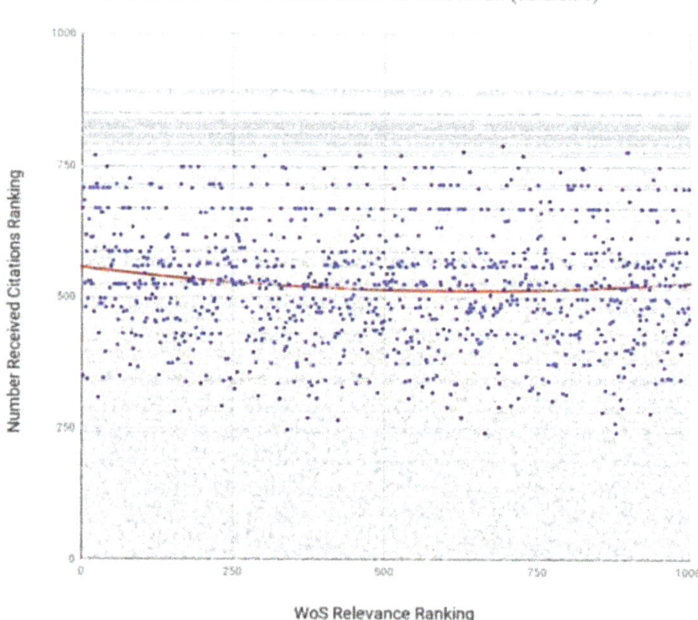

Figure 9. WoS Searches with Two Words and Title Restriction (Version 1) (rho −0.075, median in blue, the rest of data in gray).

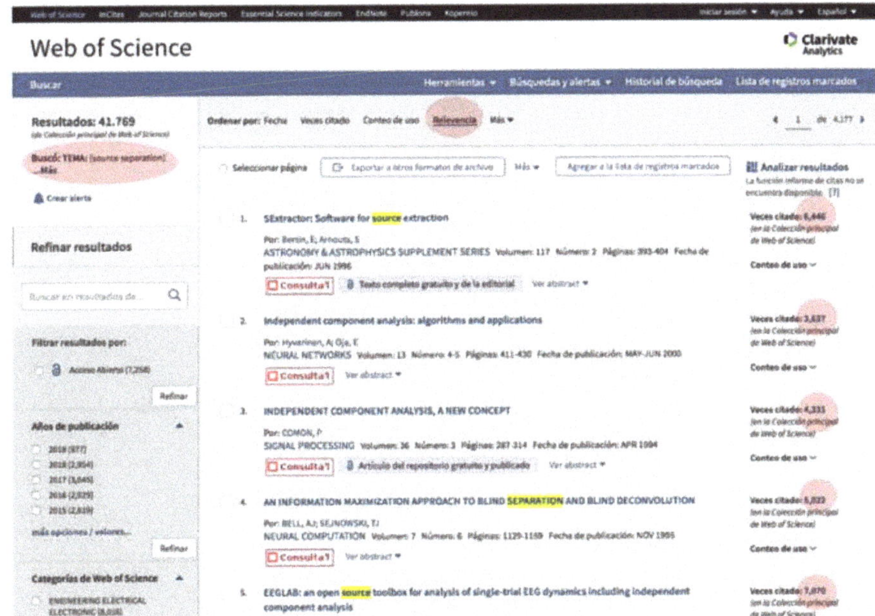

Figure 10. Search conducted on WoS with relevance ranking using the number of citations.

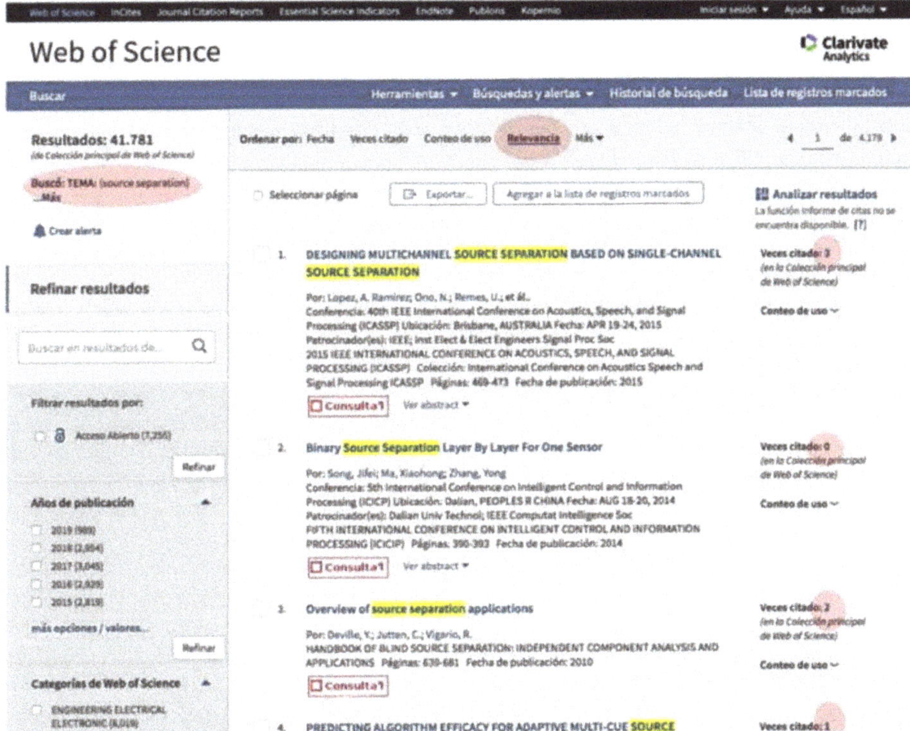

Figure 11. Same search as in Figure 10 with relevance ranking but without using the number of citations.

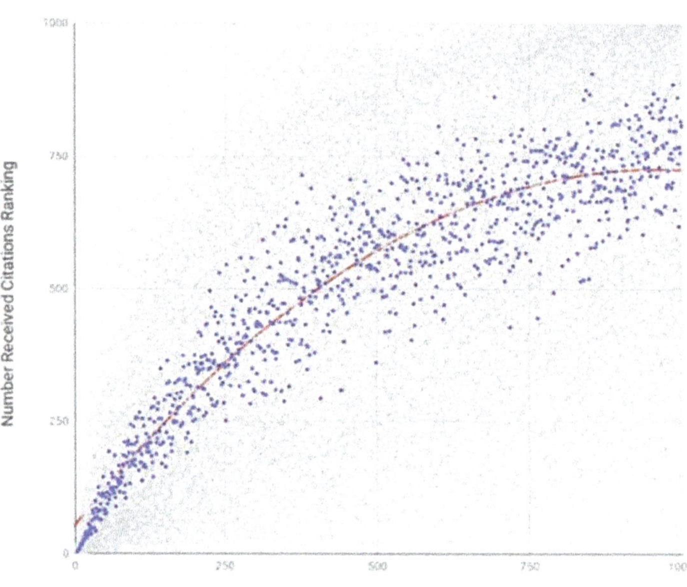

Figure 12. WoS Searches with Two Words and Title Restriction (Version 2) (rho 0.907, median in blue, the rest of data in gray).

5. Discussion

The importance attached to citations received in Google Scholar ranking of results of searches is not exactly a new finding. Beel and Gipp [31–34] described the great importance of this factor, both in full-text searches and searches by title only. However, our study incorporates methodological improvements on these earlier studies giving greater consistency to our results. Beel and Gipp applied a very basic statistical treatment, drawing conclusions from an analysis of scatter plots but without calculating correlation coefficients or conducting other specific statistical tests. Moreover, to obtain a global value of various rankings the authors took the mean. It is our understanding that the more appropriate measure of central tendency for ordinal variables is the median. Finally, the words used by the authors when conducting their searches were randomly selected from an initial list. This procedure generated a number of problems since many searches did not generate any results. In contrast, the procedure applied in the study described here is based on the most frequent words in academic documents [77–79] and the searches suggested by the academic search engines themselves based on an analysis of a large volume of user searches. This procedure ensures the random selection of the searched content and that in the vast majority of searches there are at least 1000 results. Future studies need to confirm that searches providing few results apply the same ranking criteria, as would be expected.

Beel and Gipp [31,32] found that citations received were more influential in full-text searches than they were in those restricted to the title only. Their conclusions were that Google Scholar applied two slightly different algorithms depending on the search type. Our results differ on this point as we detect a greater weighting for citations in searches restricted to just the title. There is no reason, however, to believe that different algorithms are being applied, rather it would appear to be a case of the same algorithm behaving differently depending on the factors that intervene. The presence of search words in the title is a positioning factor that forms part of the algorithm. If we ensure that all the results have the search words in the title, then we cancel out this factor and the effect of citations received is very clear. On the other hand, in full-text searches this factor does intervene and, therefore, the influence of citations received is less clear since the ranking is also determined by the presence or otherwise of the words in the title.

In a study conducted by Martín-Martín et al. [33], the authors found that in Google Scholar the citations received also had a strong influence on searches by year of publication. The authors calculated Pearson's correlation coefficient and obtained values above 0.85. These results are similar to those obtained in our study. However, Martín-Martín et al. [33] adopted a somewhat unusual method for calculating the overall value of all the searches conducted, taking the arithmetic mean of the correlation coefficients. It is our understanding that to obtain a measure of central tendency shortcuts cannot be taken and it is more appropriate to obtain the median for each position and then calculate the correlation coefficient of these medians for the Google Scholar ranking.

In Rovira et al. [34], the authors focused their attention on the weight of citations received in the relevance ranking, but only in the case of Google Scholar. While in this earlier study the authors considered searches by year, author, publication and the "cited by" link, searches by keyword were not examined. However, a very similar conclusion was reached regarding citations received: namely, that they are a very important relevance ranking factor in Google Scholar. The present study has expanded this earlier work by analyzing other information retrieval systems using keyword searches, the most common search type conducted.

Finally, it is worth stressing that we have not found any previous reports on the specific criteria for the relevance ranking used by the other three systems analyzed here. As such, we believe our study provides new reliable data on these systems.

Relevance is a concept that is clearly very much open to interpretation since it seeks to identify items presenting the highest quality, a characteristic with a very strong subjective element. The diversity of algorithms for determining relevance is a clear indicator of the complexity of its automation. For this reason, citations received is granted so much weight.

6. Conclusions

Our results indicate that citation counts are probably the main factor employed by Google Scholar and Microsoft Academic in their ranking algorithms. In the case of Scopus, by contrast, we find no evidence that citations are taken into account, as indeed is reported in the database's supporting documentation [39].

In the specific case of WoS, we detected two distinct rankings. In the initial data collection exercise, the ranking of results was conducted according to the criteria described in the WoS documentation, that is, without applying citation counts and weighting the results according to the position and frequency of keywords, as Elsevier [40] states in its documentation for this service. However, somewhat surprisingly, in a second data gathering process it became evident that the ranking on this occasion was, in essence, based on citations received. It would seem that these two distinct ranking systems were detected because WoS was undertaking tests with a view to changing its algorithm and, as such, modified its ranking criteria to obtain a better understanding of user behavior.

Our findings allow us to improve the experimental foundations of ASEO and enable us to offer useful suggestions to authors as to how they might optimize the ranking of their research in the main academic information retrieval systems. Greater visibility is implicit of a greater probability of their being read and cited [61,63] and, thereby, of boosting authors' chances to improve their h-index [85]. Any information that allows us to identify the factors that intervene in relevance ranking is of great value, not so that we might manipulate the ranking results—something that is clearly undesirable—but rather so that we can take them into account when promoting the visibility of the academic production of an author or a research group.

Other academic databases are emerging, including Dimensions and Lens, but they do not provide the same coverage as the two databases considered here. Nevertheless, we cannot rule out the possibility of their being analyzed in future studies. Such studies should usefully seek to undertake the simultaneous analysis of various factors, including, for example, citations received and keywords in a document's title, as we have discussed above. One of the limitations of this study is precisely that a single factor is studied in isolation, when a ranking algorithm employs many factors simultaneously. It would be of particular interest to analyze whether such algorithms employ interactions between several factors.

Author Contributions: Conceptualization, C.R.; Methodology, C.R.; Validation L.C., F.G.-S. and C.L.; Investigation C.R., L.C. and F.G.-S.; Resources, C.R. and C.L.; Data Curation, C.R.; Writing—Original Draft Preparation, C.R.; Writing—Review and Editing, L.C., F.G.-S. and C.L.; Supervision L.C. and F.G.-S.

Funding: This research was funded by the project "Interactive storytelling and digital visibility in interactive documentary and structured journalism". RTI2018-095714-B-C21, ERDF and Ministry of Science, Innovation and Universities (Spain).

Conflicts of Interest: The authors declare no conflicts of interest.

Appendix A. List of Terms Used in The Searches

One-Term Searches

Search words obtained from [78].

Table A1. Words and Rho of one-term searches. ** $p < 0.01$.

Search Words	Rho Google Scholar Title	Rho Google Scholar Not Title	Rho Microsoft Academic Title
Median	0.990 **	0.968 **	0.907 **
approach	0.742 **	0.700 **	0.548 **
assessment	0.632 **	0.563 **	0.545 **
authority	0.868 **	0.815 **	0.465 **
consistent	0.956 **	0.783 **	0.596 **
context	0.851 **	0.681 **	0.483 **
data	0.645 **	0.662 **	0.601 **
definition	0.907 **	0.783 **	0.636 **
derived	0.905 **	0.682 **	0.568 **
distribution	0.781 **	0.649 **	0.458 **
estimate	0.939 **	0.813 **	0.527 **
evidence	0.761 **	0.616 **	0.488 **
fact	0.899 **	0.229 **	0.517 **
factor	0.490 **	0.591 **	0.490 **
formula	0.872 **	0.773 **	0.352 **
function	0.789 **	0.650 **	0.529 **
interpretation	0.852 **	0.723 **	0.570 **
method	0.762 **	0.665 **	0.613 **
percent	0.932 **	0.861 **	0.478 **
principle	0.879 **	0.812 **	0.530 **
research	0.500 **	0.642 **	0.521 **
response	0.741 **	0.603 **	0.500 **
significant	0.929 **	0.709 **	0.557 **
source	0.848 **	0.735 **	0.546 **
theory	0.488 **	0.544 **	0.483 **
variable	0.888 **	0.727 **	0.569 **

Two-Term Searches

Search words obtained from the above list and by selecting the search suggestions provided by Google Scholar and Microsoft Academic with the greatest number of results.

Table A2. Words and Rho of two-term searches. ** $p < 0.01$, * $p < 0.05$.

Search Words	Rho GS Title	Rho GS Not Title	Rho MA Title	Rho Scopus	Rho WoS Version 1	Rho WoS Version 2
Median	0.994 **	0.721 **	0.937 **	−0.107 **	−0.075*	0.907 **
approaches management	0.391 **	0.447 **	0.587 **	−0.004	−0.102 **	0.581 **
area network	0.871 **	0.108 **	0.462 **	0.025	−0.054	0.610 **
assessment learning	0.960 **	0.646 **	0.683 **	0.009	−0.038	0.605 **
assessment tool	0.855 **	0.476 **	0.619 **	−0.006	0.058	0.556 **
benefit cost	0.602 **	0.467 **	0.490 **	−0.048	−0.008	0.522 **
context awareness	0.918 **	0.302 **	0.752 **	−0.066*	−0.056	0.580 **
context model	0.956 **	−0.003	0.616 **	−0.042	0.023	0.624 **
data mining	0.875 **	0.818 **	0.747 **	0.072*	0.009	0.654 **
distribution function	0.868 **	0.171 **	0.415 **	−0.051	−0.069*	0.520 **
environment engineering	0.869 **	0.869 **	0.539 **	−0.037	−0.032	0.647 **
environment impact	0.842 **	0.120 **	0.559 **	−0.026	−0.065*	0.605 **
evidence practice	0.968 **	0.206 **	0.712 **	0.021	0.005	0.517 **
function approximation	0.903 **	0.162 **	0.619 **	0.053	−0.036	0.646 **
period time	0.956 **	0.027	0.525 **	−0.102 **	0.036	0.554 **
probability distribution	0.913 **	−0.006	0.522 **	−0.077*	−0.055	0.683 **
research design	0.658 **	0.213 **	0.648 **	0.099 **	0.063*	0.572 **
response rate	0.881 **	0.176 **	0.522 **	−0.035	0.114 **	0.606 **
response time	0.851 **	−0.072*	0.469 **	−0.016	−0.043	0.543 **
source code	0.887 **	0.082 **	0.631 **	−0.119 **	0.043	0.621 **
source separation	0.939 **	0.466 **	0.734 **	0.053	−0.062	0.588 **
statistical theory	0.772 **	0.203 **	0.573 **	−0.085 **	−0.081*	0.657 **
structure factor	0.821 **	0.334 **	0.575 **	−0.209 **	−0.026	0.607 **
structure function	0.804 **	−0.01	0.582 **	−0.208 **	−0.104 **	0.613 **
theory mind	0.962 **	0.838 **	0.680 **	−0.048	0.181 **	0.672 **
variable number	0.958 **	−0.120 **	0.656 **	0.072*	−0.001	0.081*

Appendix B. Data Files

Rovira, C.; Codina, L.; Guerrero-Solé, F.; Lopezosa, C. Data set of the article: Ranking by relevance and citation counts, a comparative study: Google Scholar, Microsoft Academic, WoS and Scopus (Version 1) (Data set). Zenodo. Available online: http://doi.org/10.5281/zenodo.3381151 (accessed on 10 September 2019).

References

1. Baeza-Yates, R.; Ribeiro-Neto, B. *Modern Information Retrieval*; Addison-Wesley Professional: New York, NY, USA, 2010; pp. 3–339.
2. Salton, G.; McGill, M.J. *Introduction to Modern Information Retrieval*; McGraw Hill: New York, NY, USA, 1987; pp. 1–400.
3. Blair, D.C. *Language and Representation in Information Retrieval*; Elsevier: Amsterdam, The Netherlands, 1990; pp. 1–350.
4. Maciá-Domene, F. *SEO: Técnicas Avanzadas*; Anaya: Barcelona, Spain, 2015; pp. 1–408.
5. Chang, Y.; Aung, Z. AuthorRank: A New Scheme for Identifying Field-Specific Key Researchers. In Proceedings of the CONF-IRM 2015, New York, NY, USA, 16 January 2015; pp. 1–13. Available online: http://aisel.aisnet.org/confirm2015/46 (accessed on 1 July 2019).
6. Brin, S.; Page, L. The anatomy of a large-scale hypertextual Web search engine. *Comput. Netw. ISDN Syst.* **1998**, *30*, 107–117. [CrossRef]
7. Kleinberg, J.M. Authoritative sources in a hyperlinked environment. *JACM* **1999**, *46*, 604–632. [CrossRef]
8. Romero, D.M.; Galuba, W.; Asur, S.; Huberman, B.A. Influence and passivity in social media. *Mach. Learn. Knowl. Discov. Databases* **2011**, *6913*, 18–33.
9. Tunkelang, D. A Twitter Analog to PageRank. The Noisy Channel Blog. 2009. Available online: http://thenoisychannel.com/ (accessed on 1 July 2019).

10. Weng, J.; Lim, E.-P.; Jiang, J.; He, Q. TwitterRank: Finding topic-sensitive influential Twitterers. In Proceedings of the 3rd ACM International Conference on Web Search and Data Mining (WSDM), New York, NY, USA, 3–6 February 2010; pp. 261–270.

11. Yamaguchi, Y.; Takahashi, T.; Amagasa, T.; Kitagawa, H. TURank: Twitter user ranking based on user-tweet graph analysis. In *Web Information Systems Engineering—WISE 2010*; Springer: Berlin/Heidelberg, Germany, 2010; Volume 6488, pp. 240–253.

12. Hristidis, V.; Hwang, H.; Papakonstantinou, Y. Authority-based keyword search in databases. *ACM Trans. Auton. Adapt. Syst.* **2008**, *33*, 11–14. [CrossRef]

13. Nie, Z.; Zhang, Y.; Wen, J.R.; Ma, W.Y. Object-level ranking: Bringing order to web objects. In Proceedings of the 14th International Conference on World Wide Web (ACM), Chiba, Japan, 10–14 May 2005; pp. 567–574. [CrossRef]

14. Chen, L.; Nayak, R. Expertise analysis in a question answer portal for author ranking 2008. In Proceedings of the IEEE/WIC/ACM International Conference on Web Intelligence and Intelligent Agent Technology (WI-IAT), Washington, DC, USA, 31 August–3 September 2010; pp. 134–140.

15. Walker, D.; Xie, H.; Yan, K.-K.; Maslov, S. Ranking scientific publications using a model of network traffic. *J. Stat. Mech. Theory Exp.* **2007**, *06*, 6–10. [CrossRef]

16. Google. How Google Search Works. Learn How Google Discovers, Crawls, and Serves Web Pages, Search Console Help. Available online: https://support.google.com/webmasters/answer/70897?hl=en (accessed on 1 July 2019).

17. Ziakis, C.; Vlachopoulou, M.; Kyrkoudis, T.; Karagkiozidou, M. Important Factors for Improving Google Search Rank. *Future Internet* **2019**, *11*, 32. [CrossRef]

18. Ratcliff, C. WebPromo's Q & A with Google's Andrey Lipattsev, Search Engine Watch. Available online: https://searchenginewatch.com/2016/04/06/webpromos-qa-with-googles-andrey-lipattsev-transcript (accessed on 1 July 2019).

19. Schwartz, B. Now We Know: Here Are Google's Top 3 Search Ranking Factors, Search Engine Land. Available online: http://searchengineland.com/now-know-googles-top-three-search-ranking-factors-245882 (accessed on 1 July 2019).

20. Beel, J.; Gipp, B. Academic search engine spam and Google Scholar's resilience against it. *J. Electron. Publ.* **2010**, *13*, 1–28. [CrossRef]

21. Enge, E.; Spencer, S.; Stricchiola, J. *The Art of SEO: Mastering Search Engine Optimization*; O'Reilly Media: Sebastopol, CA, USA; Boston, MA, USA, 2015; pp. 1–670. Available online: https://books.google.co.in/books?id=hg5iCgAAQBAJ (accessed on 1 July 2019).

22. Beel, J.; Gipp, B. Google Scholar's ranking algorithm: The impact of articles' age (an empirical study). In Proceedings of the Sixth International Conference on Information Technology: New Generations, ITNG'09, Las Vegas, NA, USA, 27–29 April 2009; pp. 160–164.

23. Beel, J.; Gipp, B.; Wilde, E. Academic search engine optimization (ASEO) optimizing scholarly literature for google scholar & co. *J. Sch. Publ.* **2010**, *41*, 176–190. [CrossRef]

24. Codina, L. SEO Académico: Definición, Componentes y Guía de Herramientas, Codina, Lluís. Available online: https://www.lluiscodina.com/seo-academico-guia/ (accessed on 1 July 2019).

25. Martín-Martín, A.; Ayllón, J.M.; Orduña-Malea, E.; López-Cózar, E.D. *Google Scholar Metrics Released: A Matter of Languages and Something Else*; EC3 Working Papers; EC3: Granada, Spain, 2016; pp. 1–14. Available online: https://arxiv.org/abs/1607.06260v1 (accessed on 1 July 2019).

26. Muñoz-Martín, B. Incrementa el impacto de tus artículos y blogs: De la invisibilidad a la visibilidad. *Rev. Soc. Otorrinolaringol. Castilla Leon Cantab. La Rioja* **2015**, *6*, 6–32. Available online: http://hdl.handle.net/10366/126907 (accessed on 1 July 2019).

27. Gielen, M.; Rosen, J. Reverse Engineering the YouTube, Tubefilter.com. Available online: http://www.tubefilter.com/2016/06/23/reverse-engineering-youtube-algorithm/ (accessed on 1 July 2019).

28. Localseoguide. Local SEO Ranking Factors Study 2016, Localseoguide. Available online: http://www.localseoguide.com/guides/2016-local-seo-ranking-factors/ (accessed on 1 July 2019).

29. Searchmetrics. Rebooting Ranking Factors. Available online: http://www.searchmetrics.com/knowledge-base/ranking-factors/ (accessed on 1 July 2019).

30. MOZ. Google Algorithm Update History. Available online: https://moz.com/google-algorithm-change (accessed on 1 July 2019).

31. Beel, J.; Gipp, B. Google scholar's ranking algorithm: An introductory overview. In Proceedings of the 12th International Conference on Scientometrics and Informetrics, ISSI'09, Istanbul, Turkey, 14–17 July 2009; pp. 230–241.

32. Beel, J.; Gipp, B. Google scholar's ranking algorithm: The impact of citation counts (an empirical study). In Proceedings of the Third International Conference on Research Challenges in Information Science, RCIS 2009c, Nice, France, 22–24 April 2009; pp. 439–446.

33. Martín-Martín, A.; Orduña-Malea, E.; Ayllón, J.M.; López-Cózar, E.D. *Does Google Scholar Contain all Highly Cited Documents (1950–2013)*; EC3 Working Papers; EC3: Granada, Spain, 2014; pp. 1–96. Available online: https://arxiv.org/abs/1410.8464 (accessed on 1 July 2019).

34. Rovira, C.; Guerrero-Solé, F.; Codina, L. Received citations as a main SEO factor of Google Scholar results ranking. *El Profesional de la Información* **2018**, *27*, 559–569. [CrossRef]

35. Thelwall, M. Does Microsoft Academic Find Early Citations? *Scientometrics* **2018**, *114*, 325–334. Available online: https://wlv.openrepository.com/bitstream/handle/2436/620806/?sequence=1 (accessed on 1 July 2019). [CrossRef]

36. Hug, S.E.; Ochsner, M.; Brändle, M.P. Citation analysis with microsoft academic. *Scientometrics* **2017**, *110*, 371–378. Available online: https://arxiv.org/pdf/1609.05354.pdf (accessed on 1 July 2019). [CrossRef]

37. Harzing, A.W.; Alakangas, S. Microsoft Academic: Is the phoenix getting wings? *Scientometrics* **2017**, *110*, 371–383. Available online: http://eprints.mdx.ac.uk/20937/1/mas2.pdf (accessed on 1 July 2019). [CrossRef]

38. Orduña-Malea, E.; Martín-Martín, A.; Ayllon, J.M.; Delgado-Lopez-Cozar, E. The silent fading of an academic search engine: The case of Microsoft Academic Search. *Online Inf. Rev.* **2014**, *38*, 936–953. Available online: https://riunet.upv.es/bitstream/handle/10251/82266/silent-fading-microsoft-academic-search.pdf?sequence=2 (accessed on 1 July 2019). [CrossRef]

39. Clarivate. Colección Principal de Web of Science Ayuda. Available online: https://images-webofknowledge-com.sare.upf.edu/WOKRS532MR24/help/es_LA/WOS/hs_sort_options.html (accessed on 1 July 2019).

40. Elsevier. Scopus: Access and Use Support Center. What Does "Relevance" Mean in Scopus? Available online: https://service.elsevier.com/app/answers/detail/a_id/14182/supporthub/scopus/ (accessed on 1 July 2019).

41. AlRyalat, S.A.; Malkawi, L.W.; Momani, S.M. Comparing Bibliometric Analysis Using PubMed, Scopus, and Web of Science Databases. *J. Vis. Exp.* 2018. Available online: https://www.jove.com/video/58494/comparing-bibliometric-analysis-using-pubmed-scopus-web-science (accessed on 1 July 2019).

42. Giustini, D.; Boulos, M.N.K. Google Scholar is not enough to be used alone for systematic reviews. *Online J. Public Health Inform.* **2013**, *5*, 1–9. [CrossRef]

43. Walters, W.H. Google Scholar search performance: Comparative recall and precision. *Portal Libr. Acad.* **2008**, *9*, 5–24. [CrossRef]

44. De-Winter, J.C.F.; Zadpoor, A.A.; Dodou, D. The expansion of Google Scholar versus Web of Science: A longitudinal study. *Scientometrics* **2014**, *98*, 1547–1565. [CrossRef]

45. Harzing,, A.-W. A preliminary test of Google Scholar as a source for citation data: A longitudinal study of Nobel prize winners. *Scientometrics* **2013**, *94*, 1057–1075. [CrossRef]

46. Harzing,, A.-W. A longitudinal study of Google Scholar coverage between 2012 and 2013. *Scientometrics* **2014**, *98*, 565–575. [CrossRef]

47. De-Groote, S.L.; Raszewski, R. Coverage of Google Scholar, Scopus, and Web of Science: A casestudy of the h-index in nursing. *Nurs. Outlook* **2012**, *60*, 391–400. [CrossRef]

48. Orduña-Malea, E.; Ayllón, J.-M.; Martín-Martín, A.; Delgado-López-Cózar, E. *About the Size of Google Scholar: Playing the Numbers*; EC3 Working Papers; EC3: Granada, Spain, 2014; Available online: https://arxiv.org/abs/1407.6239 (accessed on 1 July 2019).

49. Orduña-Malea, E.; Ayllón, J.-M.; Martín-Martín, A.; Delgado-López-Cózar, E. Methods for estimating the size of Google Scholar. *Scientometrics* **2015**, *104*, 931–949. [CrossRef]

50. Pedersen, L.A.; Arendt, J. Decrease in free computer science papers found through Google Scholar. *Online Inf. Rev.* **2014**, *38*, 348–361. [CrossRef]

51. Jamali, H.R.; Nabavi, M. Open access and sources of full-text articles in Google Scholar in different subject fields. *Scientometrics* **2015**, *105*, 1635–1651. [CrossRef]

52. Van-Aalst, J. Using Google Scholar to estimate the impact of journal articles in education. *Educ. Res.* **2010**, *39*, 387–400. Available online: https://goo.gl/p1mDBi (accessed on 1 July 2019). [CrossRef]

53. Jacsó, P. Testing the calculation of a realistic h-index in Google Scholar, Scopus, and Web of Science for FW Lancaster. *Libr. Trends* **2008**, *56*, 784–815. [CrossRef]
54. Jacsó, P. The pros and cons of computing the h-index using Google Scholar. *Online Inf. Rev.* **2008**, *32*, 437–452. [CrossRef]
55. Jacsó, P. Calculating the h-index and other bibliometric and scientometric indicators from Google Scholar with the Publish or Perish software. *Online Inf. Rev.* **2009**, *33*, 1189–1200. [CrossRef]
56. Jacsó, P. Using Google Scholar for journal impact factors and the h-index in nationwide publishing assessments in academia—Siren songs and air-raid sirens. *Online Inf. Rev.* **2012**, *36*, 462–478. [CrossRef]
57. Martín-Martín, A.; Orduña-Malea, E.; Harzing, A.-W.; Delgado-López-Cózar, E. Can we use Google Scholar to identify highly-cited documents? *J. Informetr.* **2017**, *11*, 152–163. [CrossRef]
58. Aguillo, I.F. Is Google Scholar useful for bibliometrics? A webometric analysis. *Scientometrics* **2012**, *91*, 343–351. Available online: https://goo.gl/nYBmZb (accessed on 1 July 2019). [CrossRef]
59. Delgado-López-Cózar, E.; Robinson-García, N.; Torres-Salinas, D. *Manipular Google Scholar Citations y Google Scholar Metrics: Simple, Sencillo y Tentador*; EC3 working papers; Universidad De Granada: Granada, Spain, 2012; Available online: http://hdl.handle.net/10481/20469 (accessed on 1 July 2019).
60. Delgado-López-Cózar, E.; Robinson-García, N.; Torres-Salinas, D. The Google Scholar experiment: How to index false papers and manipulate bibliometric indicators. *J. Assoc. Inf. Sci. Technol.* **2014**, *65*, 446–454. [CrossRef]
61. Martín-Martín, A.; Orduña-Malea, E.; Ayllón, J.-M.; Delgado-López-Cózar, E. Back to the past: On the shoulders of an academic search engine giant. *Scientometrics* **2016**, *107*, 1477–1487. [CrossRef]
62. Jamali, H.R.; Asadi, S. Google and the scholar: The role of Google in scientists' information-seeking behaviour. *Online Inf. Rev.* **2010**, *34*, 282–294. [CrossRef]
63. Marcos, M.-C.; González-Caro, C. Comportamiento de los usuarios en la página de resultados de los buscadores. Un estudio basado en eye tracking. *El Profesional de la Información* **2010**, *19*, 348–358. [CrossRef]
64. Torres-Salinas, D.; Ruiz-Pérez, R.; Delgado-López-Cózar, E. Google scholar como herramienta para la evaluación científica. *El Profesional de la Información* **2009**, *18*, 501–510. [CrossRef]
65. Ortega, J.L. *Academic Search Engines: A Quantitative Outlook*; Elsevier: Amsterdam, The Netherlands, 2014.
66. Hug, S.E.; Brändle, M.P. The coverage of Microsoft Academic: Analyzing the publication output of a university. *Scientometrics* **2017**, *113*, 1551–1571. [CrossRef]
67. MOZ. Search Engine Ranking Factors 2015. Available online: https://moz.com/search-ranking-factors/correlations (accessed on 1 July 2019).
68. Van der Graaf, P. Reverse Engineering Search Engine Algorithms is Getting Harder, Searchenginewatch. Available online: https://searchenginewatch.com/sew/how-to/2182553/reverse-engineering-search-engine-algorithms-getting-harder (accessed on 1 July 2019).
69. Dave, D. 11 Things You Must Know About Google's 200 Ranking Factors. Search Engine Journal. Available online: https://www.searchenginejournal.com/google-200-ranking-factors-facts/265085/ (accessed on 10 September 2019).
70. Chariton, R. Google Algorithm—What Are the 200 Variables? Available online: https://www.webmasterworld.com/google/4030020.htm (accessed on 10 September 2019).
71. Google. About Google Scholar. Available online: http://scholar.google.com/intl/en/scholar/about.html (accessed on 1 July 2019).
72. Mayr, P.; Walter, A.-K. An exploratory study of google scholar. *Online Inf. Rev.* **2007**, *31*, 814–830. [CrossRef]
73. Sinha, A.; Shen, Z.; Song, Y.; Ma, H.; Eide, D.; Hsu, B.; Wang, K. An Overview of Microsoft Academic Service (MAS) and Applications. In Proceedings of the 24th International Conference on World Wide Web (WWW 2015 Companion), ACM, New York, NY, USA, 18–22 May 2015; pp. 243–246. [CrossRef]
74. Herrmannova, H.; Knoth., P. An Analysis of the Microsoft Academic Graph. *D Lib Mag.* **2016**, *22*, 9–10. [CrossRef]
75. Ortega, J.L. *Microsoft Academic Search: The multi-object engine. Academic Search Engines: A Quantitative Outlook*; Ortega, J.L., Ed.; Elsevier: Oxford, UK, 2014; pp. 71–108.
76. Microsoft Academic. How Is MA Different from other Academic Search Engines? 2019. Available online: https://academic.microsoft.com/faq (accessed on 1 July 2019).
77. Eli, P. Academic Word List words (Coxhead, 2000). Available online: https://www.vocabulary.com/lists/218701 (accessed on 1 July 2019).

78. Wiktionary: Academic word list. Available online: https://simple.wiktionary.org/wiki/Wiktionary:Academic_word_list (accessed on 18 September 2019).

79. Coxhead, A. A new academic word list. *TESOL Q.* **2012**, *34*, 213–238. [CrossRef]

80. Harzing, A.-W. Publish or Perish. Available online: https://harzing.com/resources/publish-or-perish (accessed on 1 July 2019).

81. Harzing, A.W. *The Publish or Perish Book: Your Guide to Effective and Responsible Citation Analysis*; Tarma Software Research Pty Ltd: Melbourne, Australia, 2011; pp. 339–342. Available online: https://EconPapers.repec.org/RePEc:spr:scient:v:88:y:2011:i:1:d:10.1007_s11192$-$011$-$0388-8 (accessed on 1 July 2019).

82. R Core Team. R: A Language and Environment for Statistical Computing, R Foundation for Statistical Computing. Available online: https://www.R-project.org (accessed on 1 July 2019).

83. Revelle, W. Psych: Procedures for Personality and Psychological Research, Northwestern University. Available online: https://CRAN.R-project.org/pahttps://cran.r-project.org/package=psychckage=psych (accessed on 1 July 2019).

84. Lemon, J. Plotrix: A package in the red light district of R. *R News* **2006**, *6*, 8–12.

85. Farhadi, H.; Salehi, H.; Yunus, M.M.; Aghaei Chadegani, A.; Farhadi, M.; Fooladi, M.; Ale Ebrahim, N. Does it matter which citation tool is used to compare the h-index of a group of highly cited researchers? *Aust. J. Basic Appl. Sci.* **2013**, *7*, 198–202. Available online: https://ssrn.com/abstract=2259614 (accessed on 1 July 2019).

 *future internet*

Article

SEO inside Newsrooms: Reports from the Field

Dimitrios Giomelakis *, Christina Karypidou and Andreas Veglis

Media Informatics Lab, School of Journalism and Mass Communications, Aristotle University of Thessaloniki, 54625 Thessaloniki, Greece; ckarypid@jour.auth.gr (C.K.); veglis@jour.auth.gr (A.V.)
* Correspondence: dgiomela@gmail.com or dgiomela@jour.auth.gr

Received: 13 November 2019; Accepted: 10 December 2019; Published: 13 December 2019

Abstract: The journalism profession has changed dramatically in the digital age as the internet, and new technologies, in general, have created new working conditions in the media environment. Concurrently, journalists and media professionals need to be aware and possess a new set of skills connected to web technologies, as well as respond to new reading tendencies and information consumption habits. A number of studies have shown that search engines are an important source of the traffic to news websites around the world, identifying the significance of high rankings in search results. Journalists are writing to be read, and that means ensuring that their news content is found, also, by search engines. In this context, this paper represents an exploratory study on the use of search engine optimization (SEO) in news websites. A series of semi-structured, in-depth interviews with professionals at four Greek media organizations uncover trends and address issues, such as how SEO policy is operationalized and applied inside newsrooms, which are the most common optimization practices, as well as the impact on journalism and news content. Today, news publishers have embraced the use of SEO practices, something that is clear also from this study. However, the absence of a distinct SEO culture was evident in newsrooms under study. Finally, according to results, SEO strategy seems to depend on factors, such as ownership and market orientation, editorial priorities or organizational structures.

Keywords: search engine optimization; SEO; search engines; search; online journalism; media websites; news content; news articles

---

1. Introduction

Major search engines today are considered to be one of the most trusted and common services to retrieve information from the internet and at the same time the main method used for navigation for hundreds of millions of users around the world [1,2]. In this context, recent studies have indicated that a significant percentage of users turn to search engines first when shopping online or when information gathering really matters [3,4]. As a result, online search remains one of the best traffic sources for any website [5–7]; however, it should be noted that the vast majority of all search traffic comes from the top rankings in search engine results [8,9].

Web technologies, new reading tendencies and new information consumption habits create new working conditions for both media organizations and journalists in order to improve online news websites and make them more readable. On the one hand, the abundance of news media websites requires news organizations to be on all major platforms, all the time [10]. On the other hand, many people look for specific information, and their priorities area convenience, rapid access, and accuracy [11]. As internet technologies have brought about new modes of producing and consuming news content, many changes are observed in basic journalistic work processes, such as newsgathering, news production and distribution, and the way people consume news [12–17]. The journalistic profession is changing. Journalists and news media organizations are required to adapt to the new conditions, to become competitive and respond to the needs of the market by making the most of their resources [14,16,18]. News websites, online radio and web TV, are the main areas of action. It is generally observed that the arrival of digital technologies has made journalistic work both easier, enabling better monitoring of economic and political organizations, and more difficult, overwhelming journalists with more information than they can handle [19].

Search technology has evolved over the last years, using more complicated algorithms and incorporating information from Web 2.0 applications in order to provide better results [20]. As new technologies continue to develop rapidly, different news sources have also emerged, including search engines, online news aggregators, social networks and citizen journalism. A number of studies have shown that search engines are an important source of the traffic to many news websites today, identifying the importance of high rankings in search results and creating a significant challenge for the digital media outlets to keep their news content at the top of the search rankings [21–27]. Journalists are writing to be read, and that means ensuring that news content is found also, by search engines.

The current exploratory study is focused on the use of search engine optimization (SEO) in news websites. Specifically, it focuses on four Greek news websites with some of them being among the most recognized media outlets in Greece with high traffic volumes. The innovation of the study lies in the fact that it is one of the first research studies to investigate SEO practices in news websites, as well as in a journalism context with a different culture that has received little attention so far. Another strength pertains to the study design, including in-depth interviews with practitioners coming from four news publishers. Through a series of semi-structured interviews with SEO and media professionals, the study examines the familiarity of these news publishers with SEO practices, including common trends and practices inside their own newsrooms, and the perceived impact of SEO on journalism and news content.

2. About SEO

The practices designed to increase the visibility and traffic (visitors) that a website or a webpage receives from organic (i.e., unpaid) search engine results are referred as search engine optimization (SEO) [28–30]. SEO is connected with the creation of the first search engines in the early-90s, and it has been associated with the influence of search engine results. In general, the position and the frequency a site appears in the Search Engine Results Page (SERP) influence the number of visitors it will receive from the search engine's users.

SEO can be applied to many different websites and can target different types of search, including image—video search, local, news or academic search [31–33]. It is also very closely connected to e-Commerce websites [34]. Besides, SEO constitutes a part of Search Engine Marketing (SEM) and one of the leading and most influential activities in the field of online marketing which defines the steps taken to organically grow a site's relevancy by building links, writing strong content or submitting to search sites [28,35]. SEO and SEM strategies should be carried out in order to attract customers and clients for business-to-consumer (B2C) companies [36]. In general, a business website can be found via a search engine by an online user in two ways: Through a pay-per-click campaign (PPC) or through an organic result listing that is based essentially on SEO. Malaga [37] divides SEO practices into four major categories:

1. Keyword research/selection;
2. Search engine indexing;
3. On-page optimization;
4. Off-page optimization.

Keyword research is the main SEO task that involves finding and analyzing actual search terms/phrases people enter into search engines. This practice, (usually with help from keyword suggestion tools, such as Google's Keyword planner) gives SEO professionals a better understanding of how high the demand is for specific keywords, as well as how hard it would be to compete for those terms in the organic search engines results. Indexing is the process of attracting the search engine spiders to a website. All of the major search engines have a submission form where users could submit a website (entering the URL) for consideration. On-page optimization includes the management of all factors associated directly with a website, such as keywords, appropriate content, internal link structure, as well as html elements. It also contains page title (or HTML title tag), on-page headlines, description of web pages (or meta description tag) and URLs. Finally, off-page optimization includes all the actions made away from the website, such as link building or social signal strategy. Regarding link building, the more referrals someone has across the Web, the more search engine spiders notice and categorize their content [38]. Also, social signals may have a positive impact on websites and are considered as the new link building metric as search engines increasingly search for social signals to help the ranking of pages [39,40].

SEO has come a long way from its early days, and the search industry has seen many innovations from artificial intelligence to voice search. The latter looks like a fast-rising trend in web search, considering that a significant percentage of searches on mobile comes from voice searches [41–44]. Today, recent algorithm changes in search engines especially in Google, the world's most popular search engine, place more value on quality and content marketing, leading many experts to call these changes the new SEO [45]. As Ledford [29] notices, the search results are affected by the perceived quality of the page (indicated by a quality score) in accordance with the algorithm used, which includes a number of factors such as location, frequency of keywords, links or clickthrough rates.

3. SEO and News Websites

As people's reading habits change due to web technologies, online journalism finds itself having to chase web traffic [46]. Nowadays, there is no doubt that the internet is the main source of news preferred by many readers in order to get informed [47]. Search engines are used as a basic tool of navigation and filter for news by many people, as internet traffic depends to a great extent on them [25,48]. According to reports from the Reuters Institute for the Study of Journalism, search remains a significant gateway to the news in many countries, such as Poland, Turkey, Germany, France, Italy, the United States and Brazil [26,27]. As a result, the survival of a website is related to its visibility through web search [49].

Although search engine optimization appeared almost in parallel with the creation of the first search engines [50], SEO practices were only adopted by newsrooms within the last few years [32,51]. Many leading online media outlets (e.g., Daily Mail, Guardian, Los Angeles Times, Daily Telegraph) have employed SEO specialists in an attempt to win greater visibility and position their stories at the top of the search rankings. In the British Broadcasting Company (BBC), for example, journalists are trained in basic SEO [52–55]. Another important change concerns the implementation of a dual-headline system that is used until now. Specifically, a short one for the front page and other website indexes, and a longer SEO title with more characters/keywords which appear on the story page itself and also in search engine results [52,53]. Other noteworthy examples include the Los Angeles Times, as well as the Christian Science Monitor, where the incorporation of an SEO strategy or an SEO manager proved to be key factors leading to traffic increase [54,56].

The presence of SEO strategies is considered as an emerging production norm and practice that direct impacts journalistic workflow and creates new challenges for media professionals. Given the increase in online news and the dependence of the media outlets on digital platforms (especially in Google), news publishers constantly try to find techniques in order to improve the prominence and visibility of their stories in search engines and other news aggregators [6,32,51,57,58]. Today, the digital success of news organizations depends, among other things, on operational changes in the processing and distribution of news. In this context, news publishers need to consider search engine positioning strategies and implement SEO actions in their newsrooms [6,32]. Application of SEO practices to digital media outlets can be divided into three broad categories: On-page, off-page and technical SEO. Investing in the appropriate systems and training, news organizations need to alter their content to attract the interest of the bots and thus, improve the exposure of their news stories in search engines [54]. Regarding on-page SEO, many news publishers today around the world create news rich in keywords: They create SEO-friendly titles, they use metadata, they include relevant keywords in the initial paragraphs (synonyms, plural variations or other forms) and use multimedia content in the form of videos, photographs, podcasts, etc. [6,32]. In contrast, off-page SEO refers to all the actions carried out off the web page, such as obtain as many quality incoming links as possible or disseminate the news content on the social platforms. Finally, technical SEO may include actions like the use of special formats for the mobile web, good information architecture or an increase in website speed [32,57,58].

Today, SEO is considered among the key journalistic skills that a modern journalist must possess. However, the development of online journalism is accompanied by a significant dependence of news publishers on technology firms that run the function of infomediation (a mix of edition, aggregation and distribution of third-party content that connects information supply with information demand). This function has changed news production practices and also it is creating conflicts between journalistic values, norms and new digital practices [6]. The entrance of SEO strategies in news organizations was—and is sometimes still—criticized by some people who believe that they downgrade the journalistic work. This belief derives from the observation that journalists change their news agenda and the way they write, producing content written mainly for machines. According to Giomelakis and Veglis [6], the aim of SEO is not better or more diverse journalism. However, journalistic work can be benefited if it is implemented consciously and wisely. SEO does not require media professionals to dumb down or writing with the only purpose to achieve better rankings. Besides, journalists are still writing to be read, and these practices can help their articles to be found [6,59]. Quality content made from professionals is still necessary, and thus, the role of the journalist is more important than SEO. During the writing process, a web editor must feel creative, combining SEO with quality content production. Given the wealth of news websites and the rapid dissemination of information, the main goal is a story to be found by readers also through search engines and news aggregators [59]. This means that journalists, and media organizations in general, must adapt to the new circumstances. As the world of SEO continues to evolve, SEO workers in every business must be up to date on new developments and also utilize useful tools and services for their work.

4. Methodology

This study is focused on the application of SEO practices inside newsrooms and news media outlets. The paper draws on data derived from in-depth interviews with practitioners coming from four news publishers in Greece with some of them being among the most recognized media outlets with high traffic volumes. The only prerequisite was the respondents and media representatives to have a position of SEO manager or be responsible for SEO strategy in every newsroom. The sample of four interviewees included an SEO manager, two general directors and one business owner. The main goal of the study was to examine different types of media organizations with different characteristics. Thus, the sample (see Table 1) included some long-established online news publishers along with a newer media outlet (with online presences ranging from 3 to 12 years), both nationwide and local media, and finally, online-only media organizations, as well as outlets co-published with print editions.

Table 1. Greek media outlets under study.

Name	Alexa Rank (GR) *	Years	Incoming Links *	Type
Thestival.gr	60	8	1.165	Online only
Tvxs.gr	132	11	2.547	Online only
Neolaia.gr	573	12	520	Online only
Thessnews.gr	457	3	228	Newspaper/Online

* Data are based on information from the Alexa ranking system that includes top sites from all categories in Greece, not only news websites (foreign websites are also included).

Following the initial acceptance of media representatives, the interview questions and a link to an online, semi-structured questionnaire were sent electronically. Semi-structured interviews were used in order to allow an in-depth exploration of participants' responses and space to express their experiences and the trends in their newsrooms. A standard set of questions was covered, allowing flexibility for any follow-up questions and to explore other issues of relevance to participants. The interviews and data collection took place during a two-month period (from June until July 2019). The research, along with the development of the questionnaire, was based on the literature review [6,32,51,53,57–60] and was adapted in the context of the Greek media landscape. The interviews included a mixture of open-ended and closed-ended questions in order to answer the research questions (RQ) based on what previous, recent studies have found. It should be noted that the majority of the questions were open-ended, allowing the respondents to answer in their own words in open text format to better capture their complete knowledge, feeling, and understanding of the topics. Additionally, the small number of close-ended questions (including questions with a five-level Likert scale) focused on some general characteristics for every media outlet (e.g., job position of respondents, type of media outlets) and questions where respondents were asked to give their personal opinion. Regarding the latter, the respondents had the opportunity to justify their answers providing more details. This type of question (close-ended) was chosen because it was easier and quicker for respondents to answer, decreasing the likelihood of irrelevant or confused answers. Also, it was taken into account that if respondents were struggling to understand particular questions, they could read the answer options for further context (e.g., names of common SEO tools). In this case, an 'Other' answer option was added if a respondent wanted to provide a unique answer. Where it was necessary, a follow-up via telephone was carried out in order to make clear the answers or discuss other issues of relevance. Apart from some general questions about media outlets under study, questions covered many areas, including how SEO policy is operationalized and applied inside newsrooms, how long these policies have been in place and also the impact on journalism and news content. The main questions during the interviews are depicted in Table 2, according to their type (open/close-ended).

Table 2. Main questions under study. SEO, search engine optimization.

Questions	Type
How familiar are you with SEO, and how long have you been dealing with these practices?	Closed-ended (Likert scale)/open-ended
Is there any SEO expert in your organization? If not, is there another person involved in such practices?	Open-ended
What factors do you think may influence SEO usage?	Open-ended
How SEO strategy is applied inside your newsroom and which are the most common practices—tools?	Close-ended with a list of possible answers
Do you use SEO practices when writing news articles?	Open-ended
What (if any) are the consequences of SEO on news language?	Open-ended
Do you track online traffic to your site and what kind of metrics are you most interested in?	Open-ended/close-ended (Likert scale)
What impact do you think SEO has on news agenda and several publication practices?	Open-ended
How much do you think SEO improves the relationship with your audience?	Open-ended
Do you think SEO benefits journalism, in general, (or it may impact negatively)?	Open-ended

Prior to performing the main research, a pilot study was conducted in order to find any problems and discrepancies that the questions might have included. A number of improvements were made on the initial questionnaire, mainly in the fields of readability and usability. An introductory text informed the respondents about the use of the obtained information and the voluntary nature of this research. Thematic analysis (TA) using Braun and Clarke's six-phase framework [61] was used to make sense of the data. The analysis focused on examining patterns within qualitative data that are important or interesting. Specifically, it was used to explore questions about respondents' experiences, perspectives, practices or factors that influence and shape particular phenomena. The different stages that were followed in the process of analysis were:

1. Familiarizing with data

Repeated reading in order for researchers to feel familiar with the data. Also, the follow up interviews made via telephone were transcribed. In parallel, thoughts or ideas for coding and meanings were noted, generating an initial list of interesting segments across the data set.

2. Assign preliminary codes to the data

Identify interesting elements of the data (coding manually) that can be assessed in a meaningful way. Important ideas/codes providing a description of respondent's experience were identified and written in the form of a list.

3. Searching for repeated patterns and themes

When all interviews were coded separately, the different codes were sorted into potential themes. Also, the researchers considered how different codes may combine to form a primary theme.

4. Reviewing themes

Themes were reviewed and refined—coherent patterns were formed in the context of the data set. The researchers re-read data several times and went backwards and forwards between raw data, codes and themes, until they felt confident enough that different codes, could collate together and form a theme.

5. Defining and naming themes

The scope and content of each theme and sub-theme clearly defined. The researchers considered the themes themselves, and each theme in relation to the others. Different dimensions of the same pattern-phenomenon were shown at this stage.

6. Producing the report

Final analysis of repeated patterns (themes)—connection with research questions and relevant literature.

Finally, the thematic analysis identified five major themes:

- Awareness and use of SEO;
- Factors affecting the use of SEO;
- Common SEO tools/practices;
- Monitoring web traffic;
- Impact of SEO on news content creation and journalism.

5. Research Questions

Following the previous discussion and guided by the key concepts of field theory, this study attempts to answer the following research questions:

RQ1: Do the media representatives know and thus, utilize SEO practices in their working organization, and what factors may affect their use according to their point of view?

RQ2: Is there an individual SEO job position inside the newsrooms under study, and which are the most common SEO tools and practices?

RQ3: Do media outlets under study monitor their web traffic and what metrics do they focus on?

RQ4: What is the impact of SEO on news content creation inside these newsrooms?

RQ5: According to respondents' point of view, what is the impact of SEO on the journalism profession, in general, regarding the publication/selection of news, as well as the relationship with the audience?

6. Results and Discussion

6.1. Awareness and Usage of SEO (RQ1)

The respondents were questioned about their experience in using SEO practices with the answers ranged initially in five categories ((1) Poor, (2) Fair, (3) Good, (4) Very Good, (5) Excellent). Subsequently, the respondents had the opportunity to justify their answers providing more details through open-ended questions (e.g., the years they deal with SEO). Generally, the sample reported a high degree of SEO practice awareness. The majority of the respondents had excellent knowledge of them, and one of them had very good knowledge. It is very interesting that all of them have dealt with SEO for many years in their working organization, as two of them had ten-years of experience, one of them had eight-years, and the other had five-years. The result seems to be reasonable given that many studies have shown that search engines are an important source of the traffic to many news websites around the world, identifying the importance of high positions in search results [21–27]. Also, newsrooms are being increasingly asked to cater to audience interests in order to generate more traffic and "clicks" [62] and in this context many Greek media professionals have begun to emphasize more on what audiences want to know [63].

Regarding the factors that may influence SEO usage, all the respondents agreed that the use of SEO practices is more prevalent in large news organizations, which try to ensure high traffic numbers, in contrast with smaller media outlets. Moreover, they reported that ownership and market orientation (i.e., public or private), as well as the type of media organization (print, radio, television or online), are highly related to how search engine optimization is used in newsrooms. For example, private news organizations seem to use more of their resources in SEO strategy than the public ones. Also, its use might vary if media organizations are online-only or coexist with other distributions (e.g., print/broadcast channel). All the above are in agreement with prior studies where SEO strategy was found to be varied among newsrooms and factors, such as market orientation, or the ownership/business model were proved to play an important role on how new technologies are incorporated into the journalism profession [6,53,64–67]. In short, it may be concluded that news organizations tend to develop distinct forms of SEO use aligned with their organizational imperatives, structures, business model, as well as their editorial priorities.

6.2. SEO Job Position and Most Common SEO Practices/Tools (RQ2)

According to the results of this research, it is noticeable that there was no specific SEO specialist jobs in any of the news websites that were examined. The respondents referred different reasons for this, each one for the website in which they work: It may be a decision of the ownership, there may be financial reasons, or someone else works on it. Considering there were no SEO specialists working for the news websites, respondents answered that SEO is a duty of the director or the chief editor or the journalists or a freelancer. The results come in contrast with many leading media outlets across the US and other European newsrooms (e.g., Daily Mail, Guardian, BBC, Los Angeles Times) which created SEO specialist/editor positions within their newsrooms during the last years [52–55]. In this context, the incorporation of an SEO chief and SEO strategy proved to be key factors leading to traffic increase in news publishers, such as Los Angeles Times and the Christian Science Monitor [54,56].

Apart from the SEO job position, respondents were also asked to indicate the most common search engine optimization practices used in their newsrooms. According to the answers received from media representatives, the most common practices utilized by media professionals were keyword research, as well as research about hot topics, the top search queries and the users' preferences. Correspondingly, the most popular tools were Google keyword planner (for keyword research) and Google trends (that analyzes the interest and the popularity of a search term and track "buzz" online). Alexa.com services and Google search console were less popular among newsrooms. Furthermore, half of the respondents answered that backlink checker for monitoring inbound links, as well as software solutions for general SEO analysis, are often used in their newsrooms.

Regarding off-SEO and link-building strategies, all the media outlets under study tend to share their news content on social media after publishing it. All the newsrooms share their content on Facebook, three of them on Twitter and one on Instagram (all three comprising the most popular social networks in Greece), in an effort to increase the shelf life and distributed reach of quality content. The newsrooms seem to realize the significance of social media platforms in a changing news media landscape having an active presence on Facebook, Twitter and YouTube (all of them), as well as on Instagram (3 out of 4). Particularly in Greece, the media market is characterized by high use of social media, since Greek people tend to read the news and participate by commenting or sharing content [26]. In this context, social media are considered a great source of user opinions whose structure can offer useful information for the polarity classification task [68]. While search engines are increasingly becoming more sophisticated at interpreting web content, social signals and links from social media have already started to play their role in SEO. The extent of social impact on search is evolving, which constitutes a characteristic of the SEO industry [39,50,69,70]. Today, an active presence in social media is considered a requirement for a news organization and the above may affect or have an indirect impact on SEO. For example, the more Facebook page likes or Twitter followers a media outlet has, the more social signals, such as likes or tweets its posts will generate. Video content (e.g., content on YouTube) is also often ranked higher in search engines.

Finally, it is worth mentioning that 2 out of 4 representatives admitted during the interviews to the practice of links exchange in cooperation with other web sites, especially web sites with relevant content. This finding is in line with prior work [6] where it was also found that many Greek media outlets exploit this practice, establishing contact with other sites in order to share links, to boost ranking and also, cover a wider range of topics. Reciprocal linking between two websites through their content can help in building more qualitative inbound links and also in bringing an increase in traffic. It is considered a common tactic for media websites, especially when there is a lack of staff and a number of thematic categories cannot be covered properly.

6.3. Online Traffic Reports and Main Traits (RQ3)

Initially, all the respondents reported web analytics usage for monitoring website traffic in their company and all of them mentioned the use of Google analytics. The above results were reasonable, given that Google Analytics is deemed globally as the most popular web analytics software and a leading tool for sales and marketing purposes [71,72]. Moreover, the high degree of web analytics use can be characterized as unsurprising, given the growing importance of internet metrics in the journalism profession in recent years across different types of newsrooms [67,73]. Based on the data from the interviews, in three out of the four newsrooms under study, the traffic coming from search engines was around a third or more (30–40%) and in another news outlet was between 10–20%. The results are in consonance with previous studies that have shown that a large percentage of readers get informed through search engines [21–23,26–28] and confirm that, for news sites, search remains crucially important. The respondents were also asked to give information regarding the use of web analytics, such as who is in charge and who tracks them, as well as the frequency of use. In two media outlets, the traffic reports are being checked daily; one respondent reported many times throughout a day; while another one answered monitoring on a weekly basis. Chief editors have access to these tools in all newsrooms under study, while in the majority of media outlets, the traffic reports are also monitored from journalists and the marketing department. The IT departments, as well as external partners specializing in SEO, have access in two out of four newsrooms.

Regarding the most popular metrics, newsrooms were more interested in web analytic metrics that report data for the overall website content and less about specific sections of their website or about specific articles. The interest of newsrooms in online traffic metrics (concerning a variety of indicators) was measured on a 1–5 scale, with 1 signifying not at all interested to 5 indicating highly interested. Subsequently, the respondents had the opportunity to justify their answers providing more details through open-ended questions. According to the results, it was found that newsrooms use

Web Analytics to obtain information regarding general website/content metrics. For example, they seem to be very interested in the overall data for the website traffic, such as page views, sessions and unique visitors, as well as the type of content users specifically prefer to read (popular articles). Additionally, they showed significant interest regarding the visitors' behavior and metrics, such as bounce rate/exit, average time spent on the website and new/returning users. Finally, the respondents seemed to appreciate the data for the traffic sources—channels (e.g., search engines, social media, etc.), as well as the data for the search terms that led to a website. On the other hand, newsrooms seem to be less interested regarding the demographic data of users (e.g., language, country, city, age, sex), as well as technical data, such as the most used browser (e.g., Chrome, Firefox, etc.), operating system (e.g., Windows, Mac OS), screen resolution, devices, etc. Also, they showed little interest concerning the number of comments on their articles and social media sharing (e.g., Facebook, Twitter, etc.).

6.4. SEO and News Content Creation (RQ4)

The current study examined real data and certain aspects of everyday routines according to respondents' media outlets. In general, the respondents reported that SEO affects news stories, and more specifically, it may have a significant impact on the editing of the articles and news language. Based on the results, newsrooms have seen SEO practice having a direct impact on journalistic workflow and the creation of news content incorporating techniques designed to ensure high ranking in search engine results pages.

The most popular SEO practices used by media outlets implemented in news content creation were the use of keywords in SEO-friendly titles, the use of meta description tag to summarize a web page's content, and the use of internal links that point to another webpage on the same website (see Figure 1). Other widely used practices were the presence of keywords in the main text of an article, keyword tags, image optimization (with use of keywords and small sentences on file names or alternative texts/tags), external links and also the use of multimedia content. Less common practices (only two out of four media outlets) were the use of different titles and the estimation of the title character limit.

Figure 1. Most common SEO practices in news content creation.

The above results are unsurprising. Titles are considered one of the most important on-page SEO elements with major search engines, paying a lot of attention to them. Also, the use of meta description tag is a good practice and is highly recommended because they are commonly used by Google as a snippet/preview of someone's web pages on SERPs [28,29]. The addition of relevant, video content adds value, makes content even richer, and it is often ranked higher from search engines [45]. Video content attracts and engages users more, since it is deemed to be highly shareable and having a higher clickthrough rate compared to traditional text results [50]. In addition, the optimization of images helps search engines to determine easily what the image is about, and it is considered very important, especially for image-based search engines, such as Google Images [19]. Finally, links that lead to other relevant web pages (preferably high-ranking sites) can contribute positively to SEO, especially if a site is new [6,74]. In contrast with our results, the practice of different titles is often used by several large news organizations (e.g., New York Times, BBC News, Huffington Post or Guardian) and the most common practice is the different headlines between the front page and the story page itself. The latter, appearing in search engine results, are usually more specific and include more keywords [6,32,52,53].

6.5. The General Impact of SEO on Journalism (RQ5)

Other topics addressed by this research included the impact of SEO practices on issues related to both the publication and selection of news. It also examined if these practices improve relations with the public and with the journalism profession in general. According to the respondents, the position of the article on the site and the period of time it will remain online are connected with SEO, as their responses ranged from quite to very much, while one of them indicated that there is no connection at all. About the multimedia content, there was some differentiation between the respondents: Two of them said that its presence affects SEO practices a little and two of them said a lot. According to the answers, the presence and sharing of articles on social media are also connected with SEO.

Moreover, three out of four respondents indicated that the use of SEO practices within newsrooms could greatly improve their relationship with the audience, and just only one believed that there is no connection between them. The respondents were asked to give their opinion about whether SEO, in general, affects the way journalists choose the news stories that will be published. All the respondents responded that SEO, in general, affects the news agenda and news content chosen by their media outlet considerably. From the same perspective, SEO techniques seem to prefer news stories that concern topical subjects or breaking news, while making the promotion of features or opinion pieces more difficult [59]. However, respondents also reported that SEO strategy would have a greater impact on others (third person) than themselves (first person) as regards the selection of news content. From a social perspective, this might be connected to other studies where Davison's [75] initial hypothesis of the third-person effect was studied.

Finally, all the respondents considered SEO strategy as an essential tool for editors, bloggers, media professionals and anyone, in general, who publishes newsworthy content. They seemed to appreciate the usefulness of an SEO strategy, and they believed that it could only benefit journalistic work by helping news articles to be found. The above results are consistent with previous studies where journalists and media owners in Greece were found to be able to adapt to technological progress, considering it useful for their profession [15,76]. It is remarkable that there were different opinions about the journalistic product and how it is affected by the SEO practices. Only one of the respondents thought these practices do not diminish journalistic quality, while one of them thought that SEO affects it a little, one quite a lot and one very much in agreement with media professionals who talk about a negative impact on the craft of journalism and the creation of stories exclusively for search engines.

7. Conclusions and Future Extensions

This paper examined the use of search engine optimization practices in news websites and journalism using a series of semi-structured, in-depth interviews with professionals at four Greek media organizations. The aim of this study was to identify how familiar news publishers are with SEO practices, how SEO policy is applied inside newsrooms, the most common trends and practices, as well as the impact on news content. Nowadays, newsrooms and media outlets have embraced the use of SEO practices and utilize them in order to make their content more easily available through search results, something that is clear also from this study. In the same context, the study also noted specific optimization practices that are often used by news websites, such as keyword research,

research about hot topics and the top search queries or dissemination via social media. All of the four newsrooms under study incorporate several techniques designed to ensure high ranking in search results, such as the use of keywords and SEO-friendly titles, meta description tags, internal/external links, image optimization or the use of multimedia content. Furthermore, search traffic measurement tools and Google analytics services were used across all news organizations studied. According to the respondents' viewpoints, SEO strategy seems to be varied among different newsrooms, depending on factors, such as ownership model, editorial priorities and organizational structures. Finally, SEO practices may have a considerable impact on the way journalists and media professionals choose news stories, as well as their publishing practices (e.g., position of the article on the site or the amount of time it will remain online). In the context of the above findings, it should be noted that all the news publishers of the sample seem to not have a clear structure for using SEO practices, and they have adopted a more rudimentary approach utilizing different, popular off-the-shelf tools. The absence of a distinct SEO team or SEO chief is evident in the studied newsrooms where journalists or media professionals with many other responsibilities often deal with these practices.

In a constantly changing and competitive media environment, no one can take their readership for granted. While the media industry adapts to the digital age and competition increases, effective use of SEO within newsrooms seems to be an important element for attracting more online readers. SEO is not only about visibility on search engines—it also includes making a website more user-friendly. As Richmond notes [59], everything has to do with the editorial choices. SEO per se is value-neutral, and it does not require journalists to dumb down or write solely for gaining traffic. SEO practices reflect the essential needs of web users to find information, while also securing long-term promotion for journalism. The content is still the most important thing for any website, and it has to be characterized by reliability, interest and quality [6]. In this way, readers will be satisfied and motivated to share it through social networks, blog posts or forums, which is what search engines look for. Under these conditions, a journalist must be creative, but also strategic, combining SEO with quality content production.

This study is not without limitations. Firstly, as being an exploratory study, the sample included only a small number of Greek newsrooms, and thus, it is not claiming to be representative of the entire population of newsrooms and media outlets. It is reasonable that a larger sample could yield better results. Moreover, our results are dependent on the accuracy and honesty of the respondents answering the questions. Nevertheless, the main strength of the study was that it attempted to explore the role of SEO practices in media websites, as well as in a journalism context with a culture (i.e., Greek newsrooms) that has received little attention so far. Even though SEO is widely used by marketing practitioners, there is a relatively small amount of academic research that systematically attempts to capture this phenomenon and its impact on various industries. To the best of our knowledge, there is a scarcity of academic research examining the relationship between SEO and journalism. We believe that this study provides useful insights concerning the use of SEO inside newsrooms and it will open the door for further longitudinal analysis. Future extension of this work will include the repetition of the study with a larger sample size and a more varied selection of newsrooms. In this context, comparative studies with foreign online media would be another study goal in the near future.

Author Contributions: Conceptualization, D.G. and C.K.; methodology, D.G.; validation, D.G.; formal analysis, D.G.; investigation, D.G. and C.K.; resources, D.G. and C.K.; data curation, D.G.; writing—original draft preparation, D.G. and C.K.; writing—review and editing, D.G., C.K. and A.V.; visualization, D.G.; supervision, D.G. and A.V.; project administration, D.G.

Funding: This research received no external funding.

Acknowledgments: The authors would like to thank interviewees for their time, as well as for sharing their insights and a valuable overview of the SEO strategy in their news organizations.

Conflicts of Interest: The authors declare no conflict of interest.

References

1. Dod, R. How Social Media Helps SEO [Final Answer]. Search Engine Journal: Boca Raton, USA **2017**. Available online: https://www.searchenginejournal.com/social-media-seo/196185/ (accessed on 22 August 2019).
2. Purcell, K. *Search and Email Still Top the List of Most Popular Online Activities*; Pew Research Center: Washington, DC, USA, 2011. Available online: https://www.pewinternet.org/2011/08/09/search-and-email-still-top-the-list-of-most-popular-online-activities/ (accessed on 20 September 2019).
3. Purcell, K.; Brenner, J.; Rainie, L. *Search Engine Use 2012*; Pew Research Center: Washington, DC, USA, 2012. Available online: http://www.pewinternet.org/2012/03/09/search-engine-use-2012/ (accessed on 20 September 2019).
4. Murga, G. Amazon Takes 49 Percent of Consumers' First Product Search, But Search Engines Rebound, 2017. Available online: https://blog.survata.com/amazon-takes-49-percent-of-consumers-first-product-search-but-search-engines-rebound (accessed on 20 September 2019).
5. Schwartz, E. *Search Engines Still Dominate Over Social Media, Even for Millennials*; Search Engine Land: Redding, CT, USA, 2018. Available online: https://searchengineland.com/search-engines-still-dominate-over-social-media-even-for-millennials-308135 (accessed on 20 September 2019).
6. Safran, N. Natural Web Site Traffic Accounts for Nearly Half of All Traffic [Data]. 2013. Available online: http://www.conductor.com/blog/2013/06/web-traffic-natural-search-data/ (accessed on 20 September 2019).
7. Giomelakis, D.; Veglis, A. Investigating Search Engine Optimization Features in Media Websites: The Case of Greece. *Digit. J.* **2016**, *4*, 379–400.
8. Brightedge. Organic Search Is Still the Largest Channel. 2018. Available online: https://www.brightedge.com/resources/research-reports/organic-search-still-largest-channel-2017 (accessed on 20 September 2019).
9. Chitika Online Advertising Network. The Value of Google Result Positioning, 2013. Available online: http://info.chitika.com/uploads/4/9/2/1/49215843/chitikainsights-valueofgoogleresultspositioning.pdf (accessed on 18 August 2019).
10. Petrescu, P. *Google Organic Click-Through Rates in 2014*; Moz: Seattle, WA, USA, 2014. Available online: https://moz.com/blog/google-organic-click-through-rates-in-2014 (accessed on 20 September 2019).
11. Picard, R. Blogs, Tweets, Social Media and the News Business. In *Let's talk: Journalism and Social Media*, 1st ed.; Ludtke, M., Ed.; Nieman Reports: Cambridge, MA, USA, 2009; Volume 63, pp. 10–12. Available online: https://niemanreports.org/wp-content/uploads/2014/03/Fall2009.pdf (accessed on 22 August 2019).
12. Giussani, B. A New Media Tells Different Stories. *First Monday* **1997**, *2*, 4–7. Available online: http://firstmonday.org/ojs/index.php/fm/article/view/521/442 (accessed on 22 August 2019). [CrossRef]
13. Chung, S.D. Interactive features of online newspapers: Identifying patterns and predicting use of engaged readers. *J. Comput. Mediat. Commun.* **2008**, *13*, 658–679. [CrossRef]
14. Spyridou, P.; Veglis, A. Exploring structural interactivity in online newspapers: A look at the Greek Web landscape. *First Monday* **2008**, *13*, 5. Available online: http://firstmonday.org/ojs/index.php/fm/article/view/2164/1960#17 (accessed on 22 August 2019).
15. Bunce, M. Management and Resistance in the Digital Newsroom. *Journalism* **2017**, *20*, 890–905. [CrossRef]
16. Doudaki, V.; Spyridou, L.P. News Content Online: Patterns and Norms under Convergence Dynamics. *Journalism* **2015**, *16*, 257–277. [CrossRef]
17. Spyridou, P.; Matsiola, M.; Veglis, A.; Kalliris, G.; Dimoulas, C. Journalism in a State of Flux: Journalists as Agents of Technology Innovation and Emerging News Practices. *Int. Commun. Gaz.* **2013**, *75*, 76–98. [CrossRef]
18. Welbers, K.; van Atteveldt, W.; Kleinnijenhuis, J.; Ruigrok, N.; Schaper, J. News Selection Criteria in the Digital Age: Professional Norms Versus Online Audience Metrics. *Journalism* **2016**, *17*, 1037–1053. [CrossRef]
19. Kizys, R.; Juan, A.; Sawik, B.; Calvet, L. A Biased-Randomized Iterated Local Search Algorithm for Rich Portfolio Optimization. *Appl. Sci.* **2019**, *9*, 3509. [CrossRef]
20. Venturini, T.; Jacomy, M.; Bounegru, L.; Gray, J. Visual network exploration for data journalists. In *The Routledge Handbook to Developments in Digital Journalism Studies*; Eldridge, S., II, Franklin, B., Eds.; Routledge: London, UK; New York, NY, USA, 2019; pp. 265–283. Available online: https://papers.ssrn.com/sol3/papers.cfm?abstract_id=3043912 (accessed on 22 August 2019).

21. Zimmer, M. The Externalities of Search 2.0: The Emerging Privacy Threats when the Drive for the Perfect Search Engine meets Web 2.0. *First Monday* **2008**, *13*, 3. Available online: http://firstmonday.org/ojs/index.php/fm/article/view/2136/1944 (accessed on 22 August 2019). [CrossRef]

22. Newman, N. *Mainstream Media and the Distribution of News in the Age of Social Discovery*; Reuters Institute for the Study of Journalism, University of Oxford: Oxford, UK, 2011. Available online: https://reutersinstitute.politics.ox.ac.uk/sites/default/files/2017-11/Mainstream%20media%20and%20the%20distribution%20of%20news%20in%20the%20age%20of%20social%20discovery.pdf (accessed on 22 August 2019).

23. Dutton, W.H.; Blank, G. *Next Generation Users: The Internet in Britain*; Oxford Internet Survey 2011 report; Oxford Internet Institute, University of Oxford: Oxford, UK, 2011. Available online: http://www.oii.ox.ac.uk/news/?id=598 (accessed on 22 August 2019).

24. Olmstead, K.; Mitchell, A.; Rosenstiel, T. *Navigating News Online: Where People Go, How They Get There and What Lures Them Away*; Pew Research Center's Project for Excellence in Journalism: Washington, DC, USA, 2011. Available online: http://www.journalism.org/2011/05/09/navigating-news-online/ (accessed on 22 August 2019).

25. American Press Institute. The Personal News Cycle: How Americans Choose to Get Their News. This Research was Conducted by the Media Insight Project, 2014. Available online: http://www.americanpressinstitute.org/publications/reports/surveyresearch/how-americans-get-news/ (accessed on 22 August 2019).

26. Newman, N.; Levy, D.A.L. *Reuters Institute Digital News Report 2014—Tracking the Future of News*; Reuters Institute for the Study of Journalism, University of Oxford: Oxford, UK, 2014. Available online: https://reutersinstitute.politics.ox.ac.uk/our-research/digital-news-report-2014 (accessed on 22 August 2019).

27. Newman, N.; Fletcher, R.; Kalogeropoulos, A.; Levy, D.; Nielsen, R.K. *Reuters Institute Digital News Report 2017*; Reuters Institute for the Study of Journalism, University of Oxford: Oxford, UK, 2017. Available online: http://www.digitalnewsreport.org/survey/2017/ (accessed on 22 August 2019).

28. Newman, N.; Fletcher, R.; Kalogeropoulos, A.; Levy, D.; Nielsen, R.K. *Reuters Institute Digital News Report 2018*; Reuters Institute for the Study of Journalism, University of Oxford: Oxford, UK, 2018. Available online: http://www.digitalnewsreport.org/survey/2018/ (accessed on 22 August 2019).

29. Potts, K. *Web Design and Marketing Solutions for Business Websites*, 1st ed.; Apress: New York, NY, USA, 2007.

30. Ledford, L.J. *Search Engine Optimization Bible*, 2nd ed.; Wiley: Indianapolis, IN, USA, 2009.

31. Saura, J.R.; Palos-Sánchez, P.; Suárez, L.M.C. Understanding the Digital Marketing Environment with KPIs and Web Analytics. *Future Internet* **2017**, *9*, 76. [CrossRef]

32. Beel, J.; Gipp, B.; Wilde, E. Academic Search Engine Optimization (ASEO): Optimizing Scholarly Literature for Google Scholar and Co. *J. Sch. Publ.* **2010**, *41*, 176–190. [CrossRef]

33. Giomelakis, D.; Veglis, A. Employing Search Engine Optimization Techniques in Online News Articles. *Stud. Media Commun.* **2015**, *3*, 22–33. [CrossRef]

34. Wikipedia. Search Engine Optimization. Available online: https://en.wikipedia.org/wiki/Search_engine_optimization (accessed on 22 August 2019).

35. Malaga, R.A. The Value of Search Engine Optimization: An Action Research Project at a New E-Commerce Site. *J. Electron. Commer. Organ.* **2007**, *5*, 68–82. Available online: http://www.irma-international.org/viewtitle/3498/ (accessed on 22 August 2019). [CrossRef]

36. Nigro, H.O.; Balduzzi, L.; Cuesta, I.A.; Cvsaro, S.E.G. Knowledge Based System for Intelligent Search Engine Optimization. In *Recent Progress in Data Engineering and Internet Technology*; Ford Lumban Gaol; Springer: Berlin/Heidelberg, Germany, 2012; Volume 157, pp. 65–72. [CrossRef]

37. García, J.J.L.; Lizcano, D.; Ramos, C.M.; Matos, N. Digital Marketing Actions That Achieve a Better Attraction and Loyalty of Users: An Analytical Study. *Future Internet* **2019**, *11*, 130. [CrossRef]

38. Malaga, R.A. Worst Practices in Search Engine Optimization. *Commun. ACM* **2008**, *51*, 147–150. [CrossRef]

39. Ehrlich, S. SEO Best Practices: The Impact of Social Media on Search Engine Optimization. *Bulldog Reporter*. 25 November 2013. Available online: http://www.bulldogreporter.com/dailydog/article/thought-leaders/seo-best-practices-the-impact-of-social-media-on-search-engine-opti (accessed on 22 August 2019).

40. Rayson, S. The Social Media Optimization (SMO) of SEO: 7 Key Steps. *Social Media Today*. 20 August 2013. Available online: https://www.socialmediatoday.com/content/social-media-optimization-smo-seo-7-key-steps (accessed on 22 August 2019).

41. Frasco, S. 6 Reasons Social Media is Critical to Your SEO. *Social Media Today*. 9 November 2013. Available online: https://www.socialmediatoday.com/content/6-reasons-social-media-critical-your-seo (accessed on 22 August 2019).
42. Dean, B. The Definitive Guide to SEO in 2019. Available online: https://backlinko.com/seo-this-year (accessed on 22 August 2019).
43. Sterling, G. *Google Says 20 Percent of Mobile Queries are Voice Searches*; Search Engine Land: Redding, CT, USA, 2016. Available online: https://searchengineland.com/google-reveals-20-percent-queries-voice-queries-249917 (accessed on 22 August 2019).
44. Sentance, R. *The Continuing Rise of Voice Search and How You Can Adapt to It*; Search Engine Watch: Connecticut, USA, 2016. Available online: https://searchenginewatch.com/2016/05/31/the-continuing-rise-of-voice-search-and-how-you-can-adapt-to-it/ (accessed on 22 August 2019).
45. Ansley, A. Advanced SEO Techniques: A Mega Guide to Ranking in 2019. *Raventools*. 2019. Available online: https://raventools.com/blog/advanced-seo-techniques/ (accessed on 22 August 2019).
46. DeMers, J. SEO in 2013: 7 Surprisingly Simple Factors That Will Take The Lead. *Search Engine Watch: Connecticut, USA*. 2013. Available online: http://web.archive.org/web/20170905043128/http://www.searchenginejournal.com/seo-in-2013-7-surprisingly-simple-factors-that-will-take-the-lead/57092/ (accessed on 22 August 2019).
47. Tandoc, C.E., Jr. Journalism is twerking? How web analytics is changing the process of gatekeeping. *New Media Soc.* **2014**, *16*, 559–575. [CrossRef]
48. Pew Research Center for the People & the Press. Amid Criticism, Support for Media's 'Watchdog' Role Stands Out. 2013. Available online: http://www.people-press.org/2013/08/08/amid-criticism-support-for-medias-watchdog-role-stands-out/ (accessed on 22 August 2019).
49. Safran, N. *310 Million Visits: Nearly Half of All Web Site Traffic Comes From Natural Search (Data)*; Conductor: New York, NY, USA, 2013. Available online: http://www.conductor.com/blog/2013/06/data-310-million-visits-nearly-half-of-all-web-site-traffic-comes-from-natural-search/ (accessed on 22 August 2019).
50. Galitsky, B.; Levene, M. On the Economy of Web Links: Simulating the Exchange Process. *First Monday* **2004**, *9*, 1. Available online: http://firstmonday.org/ojs/index.php/fm/article/view/1109/1029 (accessed on 22 August 2019).
51. Dover, D. *Search Engine Optimization Secrets*, 1st ed.; Wiley: Indianapolis, IN, USA, 2011.
52. Codina, L.; Iglesias-García, M.; Pedraza, R.; García-Carretero, L. *Search Engine Optimization and Online Journalism: The SEO-WCP Framework*, 1st ed.; UPF: Barcelona, Spain, 2016.
53. Asser, M. *Search Engine Optimisation in BBC News*; BBC: London, UK, 2012. Available online: http://www.bbc.co.uk/blogs/internet/posts/search_engine_optimisation_in (accessed on 22 August 2019).
54. Dick, M. Search Engine Optimisation in UK News Production. *J. Pract.* **2011**, *5*, 462–477. [CrossRef]
55. Ellis, J. *Traffic Report: Why Pageviews and Engagement Are up at Latimes.com*; NiemanLab: Cambridge, MA, USA, 2011. Available online: https://www.niemanlab.org/2011/08/traffic-report-why-pageviews-and-engagement-are-up-at-latimes-com/ (accessed on 22 August 2019).
56. Oliver, L. *Mail Online Names Simon Schnieders as Search Strategy Head*; Journalism.Co.Uk: Brighton, UK, 2008. Available online: http://www.journalism.co.uk/news/mail-online-names-simon-schnieders-as-search-strategy-head/s2/a531917/ (accessed on 22 August 2019).
57. Groves, J.; Brown, C. Stopping the Presses: A Longitudinal Case Study of the Christian Science Monitor Transition from Print Daily to Web Always. *ISOJ: The Official Research Journal of the International Symposium on Online Journalism* **2011**, *1*, 86–128.
58. Lopezosa, C.; Orduna-Malea, E.; Perez-Montoro, M. Making Video News Visible: Identifying the Optimization Strategies of the Cybermedia on YouTube Using Web Metrics. *J. Pract.* **2019**. [CrossRef]
59. Lopezosa, C.; Codina, L.; Perez-Montoro, M. SEO and Digital News Media: Visibility of Cultural Information in Spain's Leading Newspapers. *Tripodos* **2019**, *44*, 41–61.
60. Richmond, S. How SEO is changing journalism. *Br. J. Rev.* **2008**, *19*, 51–55. [CrossRef]
61. Usher, N. *What Impact Is SEO Having on Journalists? Reports from the field*; NiemanLab: Cambridge, MA, USA, 2010. Available online: https://www.niemanlab.org/2010/09/what-impact-is-seo-having-on-journalists-reports-from-the-field/ (accessed on 22 August 2019).
62. Braun, V.; Clarke, V. Using thematic analysis in psychology. *Qual. Res. Psychol.* **2006**, *3*, 77–101. [CrossRef]

63. Tang, Y.E.; Sridhar, S.; Thorson, E.; Mantrala, M.K. The Bricks that Build the Clicks: Newsroom Investments and Newspaper Online Performance. *Int. J. Media Manag.* **2011**, *13*, 107–128. [CrossRef]

64. Touri, M.; Kostarella, I.; Theodosiadou, S. Journalism Culture and Professional Identity in Transit: Technology, Crisis and Opportunity in the Greek Media. In *Digital Technology and Journalism*, 1st ed.; Tong, J., Lo, S.H., Eds.; Palgrave Macmillan: Cham, Switzerland, 2017; pp. 115–139.

65. Boczkowski, P. The Processes of Adopting Multimedia and Interactivity in Three Online Newsrooms. *J. Commun.* **2004**, *54*, 197–213. [CrossRef]

66. Usher, N. Al Jazeera English Online: Understanding Web Metrics and News Production When a Quantified Audience Is Not a Commodified Audience. *Digit. Journal.* **2013**, *1*, 335–351. [CrossRef]

67. Ferrucci, P.; Tandoc, E.C., Jr. A Tale of Two Newsrooms: How Market Orientation Influences Web Analytics Use. In *Contemporary Research Methods and Data Analytics in the News Industry*, 1st ed.; Gibbs, W.J., McKendrick, J., Eds.; IGI Global: Hershey, PA, USA, 2015; pp. 58–76.

68. Cherubini, F.; Nielsen, R.K. *Editorial Analytics: How News Media Are Developing and Using Audience Data and Metrics*; Digital News Project; Reuters Institute for the Study of Journalism: Oxford, England, 2016. Available online: https://reutersinstitute.politics.ox.ac.uk/our-research/editorial-analytics-how-news-media-are-developing-and-using-audience-data-and-metrics (accessed on 22 August 2019).

69. Vairetti, C.; Martinez-Camara, E.; Maldonado, S.; Luzon, V.; Herrera, F. Enhancing the classification of social media opinions by optimizing the structural information. *Future Gener. Comput. Syst.* **2020**, *102*, 838–846. [CrossRef]

70. Pascale, A. *7 Legitimate Ways That Social Media Impacts SEO*; ClickZ: New York, NY, USA, 2014. Available online: https://www.clickz.com/7-legitimate-ways-that-social-media-impacts-seo/31746/ (accessed on 22 August 2019).

71. Stanhope, J.; Frankland, D.; Dickson, M. *The Forrester Wave: Web Analytics, Q4 2011*, 1st ed.; Forrester Research: Cambridge, MA, USA, 2011.

72. Kent, M.; Carr, B.; Husted, R.; Pop, R. Learning Web Analytics: A Tool for Strategic Communication. *Public Relat. Rev.* **2011**, *37*, 536–543. [CrossRef]

73. Hanusch, F. Web Analytics and the Functional Differentiation of Journalism Cultures: Individual, Organizational and Platform-specific Influences on Newswork. *Inf. Commun. Soc.* **2017**, *20*, 1571–1586. [CrossRef]

74. Dean, B. We Analyzed 1 Million Google Search Results. Here's What We Learned about SEO. *Backlinko.* 2 September 2016. Available online: https://backlinko.com/search-engine-ranking (accessed on 22 August 2019).

75. Davison, W.P. The Third-Person Effect in Communication. *Public Opin. Q.* **1983**, *47*, 1–15. [CrossRef]

76. Giannakoulopoulos, A.P.; Kodellas, S.N. The Impact of Web Information Availability in Journalism: The Case of Greek Journalists. In *Digital Utopia in the Media: From Discourses to Facts. A Balance*, 1st ed.; Masip, P., Rom, J., Eds.; Tripodos: Barcelona, Spain, 2005; Volume 2, pp. 547–560.

 *future internet*

Article

SEO Practices: A Study about the Way News Websites Allow the Users to Comment on Their News Articles

Minos-Athanasios Karyotakis [1,*], Evangelos Lamprou [2], Matina Kiourexidou [3] and Nikos Antonopoulos [2]

1   School of Communication, Hong Kong Baptist University, Hong Kong, China
2   Department of Digital Media and Communication, Ionian University, 28100 Kefalonia, Greece
3   Medical School, Aristotle University of Thessaloniki, 54 124 Thessaloniki, Greece
*   Correspondence: minosathkar@yahoo.gr

Received: 25 June 2019; Accepted: 28 August 2019; Published: 30 August 2019

Abstract: In the current media world, there is a huge debate about the importance of the visibility of a news website in order to secure its existence. Thus, search engine optimization (SEO) practices have emerged in the news media systems around the world. This study aimed to expand the current literature about the SEO practices by focusing on examining, via the walkthrough method, the ways that news companies allow the users to comment on their online news articles. The comments on the news websites are related to the notions of social influence, information diffusion, and play an essential role as a SEO practice, for instance, by providing content and engagement. The examined sample was collected by the most visited news websites' rankings of alexa.com for a global scale and for the countries Greece and Cyprus. The findings reveal that the news websites throughout the globe use similar features and ways to support the comments of the users. In the meantime, though, a high number of the news websites did not allow the users to use their social media accounts in order to comment the provided news articles, or provided multiple comment platforms. This trend goes against the SEO practices. It is believed that this finding is associated with the difficulty of the news organizations to regulate and protect themselves by the users' comments that promote, in some case harmful rhetoric and polarization.

Keywords: SEO; news websites; Greece; Cyprus; comments

1. Social Media Platforms, Social Influence, and Information Diffusion

The developments of technology, one of which is the Internet, have made a plethora of scholars focus their research on these transformations. From the 1980s there have existed, important changes in the information systems (IS) field alongside society. The daily use of these technological advancements is linked not only with sociological but also with psychological factors. Consequently, new models have emerged in order to provide a clearer picture of the way citizens consume and choose information [1].

Additionally, from the beginning of the related research, there was interest in the way social influence can affect the information distribution and impact [1]. Some studies proved that social influence could be more essential for women at the beginning of the information process. In addition, it may be essential in mandatory settings and it seemed to be more influential on old-aged workers. However, it was clear from the early stages of the research that social influences can change beliefs, ideas, behaviors, etc. towards other people [1].

Nevertheless, there are different forms of social influence and a vast literature that closely studies this theoretical concept [2–8]. According to Kelman [9], there are cases where the individuals accept or adopt an action or behavior despite not believing in it, as it is thought that there will be benefits or no consequences for him/her. In these cases, there is no distinct social effect or alteration in the behavior. In the meantime, social influence is related to information diffusion. Many studies in the field have revealed the connections between the social influence and the information diffusion [10–14]. Information diffusion and social influence have become of great importance due to the daily use of social media platforms and the operation of search engines.

Additionally, previous studies of social media platforms, such as YouTube, have demonstrated that users with similar cultural and political ideas tend to like or dislike the same cultural products [15]. For instance, Japan was one of the most characteristic examples of a country that became successful in the spread of its culture on a global scale. The anime and the manga cultural products have turned into global products. According to Otmazgin [16], this development allows Japan to have a greater influence on different state and non-state actors throughout the world. The aforementioned description regarding the influence of a country has led Nye [17] to create the theory of "soft power." According to him, "soft power" is the capability of causing individuals to adopt beliefs, results, opinions, etc., without getting paid. For a country, its "soft power" can be its policies and culture.

Nowadays, social influence and information diffusion are strongly linked with the World Wide Web, social media platforms, and social participation. For more than twenty years, scholars have tried to explain the role of the new technologies in the participation of the public in crucial events, such as demonstrations and collective actions [18]. The use of the Internet became the main research topic of several studies, as there was a need to explore in depth, its capabilities, and to try to predict what changes it would bring in the future for the fields of information, communication, and the social capital in a broader sense. One piece of landmark research proved an actually positive connection between the information distributed by the Internet and the construction of social capital. More specifically, the more people obtain information regarding political affairs via online networks, application, tools, etc., the more their participation in relation to political and social issues is increased [18].

Research regarding social influence and information diffusion seems to have been increased rapidly within the last years, as social media platforms have had to keep on growing promptly in the last decade. According to Statista [19], YouTube has almost 1.5 billion active users; Facebook had more than 2.3 billion monthly active users at the end of 2018 [20], and Twitter had, at the same period of time, 321 million active users [21]. These astounding growths of users for the aforementioned companies have influenced the way information is being distributed throughout society on a global scale. Some years ago, the access and the diffusion of information was not so easy, as there were a lot of technical and economic barriers that needed to be overcome in order to achieve the spread and communication of information throughout such a large network [22]. The maintenance of this vast connected network allows citizens to share openly their ideas and opinions without relying on face to face communication. In some cases, this advancement plays a significant role in expressing easier political or ideological ideas [18].

Even in more restricted media landscapes, such as the People's Republic of China (PRC), social media platforms seem to have an impact on the daily discussed topics of the public discourse. Schneider [23] argues that, in several cases, the online discussions bring on the government's agenda issues that were not about to be discussed. Overall, social media platforms provide a very different form of participating in public discourse. However, it has been proven on Facebook that there exist similar connections between offline and online political participation [24].

A form of social influence and information diffusion are the comments that appear on websites and social media platforms. Studies have proved that comments can have a significant impact on society and alter even beliefs or opinions [25–27] and also play an essential role as a search engine optimization (SEO) practice. SEO practice affects the prominence of specific websites, which are gaining more visibility if they have some unique techniques or characteristics, such as allowing comments on their websites [28–31]. Thus, this study tries to shed light on and extend the relevant literature of SEO by examining 656 popular news websites of the globe and the way they allow users to express their opinions on these websites. Some characteristics of commenting are identical to the social media platforms and are associated with social influence and, especially, with the information diffusion of the news websites. It is one of the first studies focusing on the field of news information and communication in relation to SEO practices. Meanwhile, it tries to raise awareness regarding the importance of comments in the field of SEO, as according to Dover and Dafforn [32], users' comments are one crucial parameter for improving the importance and the popularity of a website, due to the raw content and the engagement of the users. More specifically for the news websites [33] (p. 388–389), "Bounce rate is the percentage of single-page visits (visitors who enter the site and leave without viewing other pages or interacting within the same site). Do visitors spend time reading or "bounce" away quickly? If the bounce rate is very high, that indicates that the particular website as a whole is not very useful, does not engage the user and affects rankings negatively. In summary, reducing the bounce rate can result in more engaged visitors that continue deeper into the website." Additionally, studies are proving the importance of the comments on news websites for the users' engagement [34,35]. Despite the users' time taken up reading the comments, the users value the option of commenting [34].

The quality of the comments plays a significant role in improving the visibility of the website. For instance, Slegg argued in 2016 [28] that many websites remove the comments on their pages due to the lack of quality. On the other hand, other websites have a firm reliance on user's comments. Another important factor regarding the comments on a website is the comment system that the website uses. If the comment system is not fast in loading, then Google will not include the provided comments of the website for ranking the website, despite that not being considered SEO practice [29]. Besides, according to Schwartz [30], Google provides a better ranking to the websites which offer comments on their websites. The interaction and the engagement of the users remain on the website and not on other third-parties platforms, such as social media (e.g., Facebook, etc.). In this way, each website also builds each own community, which is another SEO technique for improving the Google ranking of a website [31].Therefore, the next section of the paper focuses on the importance of user's comments on websites in order to demonstrate the role that they play in promoting news websites [36,37]. It is worth mentioning that the influence of news media has been demonstrated in the scientific field of journalism and communication via several theories, such as Agenda-Setting, Cultivation, Gatekeeping, Framing, etc.

2. News Websites, Users' Comments, and SEO

In the contemporary world, a large number of news websites globally rely heavily on social media platforms in order to promote their news stories. The social media platforms, as was mentioned above, are crucial for the participation of the public and the social influence. Moreover, some social media platforms, such as Facebook, are embedded in the news websites at the end of the article, so the users can exploit their social media account and comment on the webpage of the published article. Thus, that has become the most common practice for the public to express its opinion regarding the examined issue. Therefore, it is no surprise that most news stories have been commented on by at least one user on a comment section. In this aspect of user participation, one study from the Pew Research Center arose, which proved that 25% of the grown-up population of the United States of America have commented at least one time in their lifetime on an online news story. Additionally, 37% of the study participants recognized the importance of the comments feature of a news website [38]. Furthermore, according to Ziegele and his colleagues [38], studies have shown that a large proportion of the comments section is associated with replies to other users' comments on the related section of the news article. These discussions have led to dynamic changes in the content published journalistic content and shaped a participatory culture.

Nevertheless, the users' comments can be often misleading and irrelevant to the content of the article. Moreover, there are comments that use harmful rhetoric and aim towards promoting polarization. At this point, it should be mentioned that there are even more traditional ways, such as the letter to the Editor, for the readers to express their opinions in relation to a discussed issue of the news organization. However, this practice is not considered to be common in the contemporary media world, as new innovative technology-based functions are preferred, so that there is not so much of a workload for the professionals of the news outlets for editing the opinions of the readers, and the interaction with the public is more diverse and dynamic. The users, via their participation, spend more time on the website and provide raw content, which sometimes can be characterized as "strong." The engagement and the more prolonged presence of a user on a website are linked with the bounce rate, which in turn is an element associated with SEO [39]. Additionally, according to Veglis and Giomelakis [40] the "strong" content is supposed to be one of the most important SEO tactics.

Despite more freedom of expression via the online platforms or features, there is a debate about the regulation of the comments when they contain inappropriate sentences, promote harm or hate, etc. Thus, the trend for some of the news websites is to use social platforms, such as Facebook, as the only way for the users to comment on the news articles. In this way, the news websites are not forced to use employees for regulating the comment section. The social media platform is held responsible for the inappropriate content. In addition, there were public statements that supported the use of Facebook for commenting, as it is believed to support and promote more insightful comments for the discussed issues [41]. This argument is associated also with the recent findings by Kalogeropoulos and his colleagues [42] (p. 1) who proved that "people with high interest in hard news are more likely to comment on news on both news sites and social media and share stores via social media (and people with high interest in any kind of news (hard or soft) are more likely to share stories via email)."

The visibility of a website is crucial for its existence throughout time. Therefore, news websites try to find new ways of gaining visibility to promote their content. In order to do so, they have to exploit the SEO and the news aggregators, like the Google Search Engine, which in 2003 represented the 75% of all searches [43]. Giomelakis and Veglis [40] (p. 23) argued also that "the top listing in Google's organic search results receives 32.5 percent of the traffic, compared to 17.6 percent for the second position and 11.4 for the third. In addition, sites listed on the first Google search results page generate 92 percent of all traffic from an average search."

More specifically, the more SEO practices a website employs the more visitors it has, as it takes a higher place in the related search engines. The SEO practices, amongst others, are supposed to be the most important techniques for being more profitable in the field of online marketing [40,44]. In the last decade, news organizations started to pay more attention to the SEO practices by employing specialists, to help them in gaining more visibility. The relationship between the news websites and the search engines has created a lot of controversies. For instance, Rupert Murdoch decided around a decade ago to remove his newspapers' websites from the related index of Google. Despite the importance of the SEO practices, the attractiveness of the content of the news websites still plays an important role in the visibility of the website. Actually, it seems to be the most crucial factor for the high visibility of a website [33]. Consequently, this is one of the main reasons why the current research paper studied, according to alexa.com ranking, the most prominent websites of the world, and from two countries (Greece and Cyprus), to see if the most popular news websites promote the commenting of the public via relevant commenting sections or features.

3. Methodology

The sample for this study was collected by the rankings of alexa.com for the most popular news websites in the globe (n = 353), Greece (n = 175), and Cyprus (n = 128). All the news websites were categorized into four different types of websites according to the methodology of Antonopoulos et al. (Figure 1) [36,45]. Subsequently, the categories were the following ones: (a) Television stations (the websites that are provided by television stations), (b) newspapers (the websites of traditional newspapers), (c) portal (the news companies that have only websites), and (d) mass media (when the news company has different platforms for distributing its news products). The evaluation of the websites was performed by the walkthrough method, which has been used in a lot of studies until today for revealing features and characteristics of websites and applications [46–49].

Based on what is known about SEO and the news websites [33,40,44], this study tries to expand the relevant literature by finding if the most important news websites of the world, Greece, and Cyprus allow their users to comment on the news articles of the websites. Through that method, it should be apparent whether they allow dynamic interaction between the users and the professionals of the field. It is noteworthy that according to Giomelakis and Veglis [33], the most important factor for the visibility of a website is the content of the news websites. Comments produce raw content on the websites, resulting in the users staying more on the websites and being more active than the users that do not comment. Further, the interactions, such as likes and shares, play a role in SEO and the popularity of a website [32,33,40]. However, there is not systematic research until today in the field of journalism and communication that focuses on investigating if the news websites of the globe follow these practices. Such research could expand the relevant literature of SEO and help the news websites to improve their SEO practices.

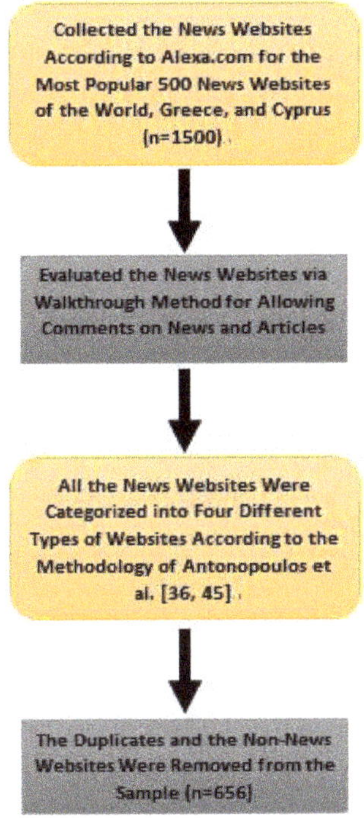

Figure 1. The summary of the methodology.

Subsequently, the following research questions were asked:

Research Question 1(RQ1): Will the majority of the news websites allow users to leave comments through social media tools and plug-ins? (Figure 2).

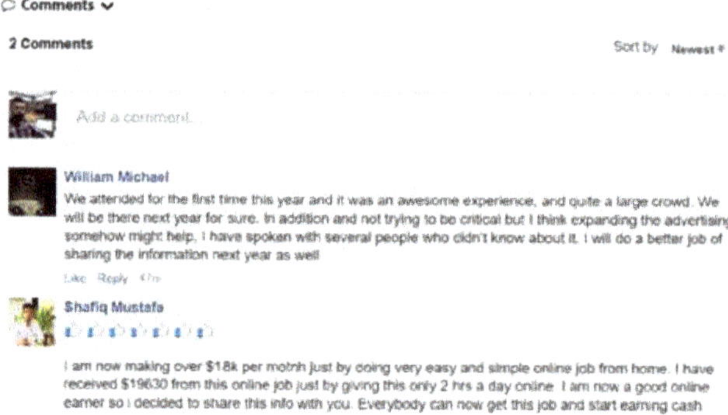

Figure 2. Example acquired from www.islandpacket.com.

Research Question 2 (RQ2): Will most of the news websites use multiple comment platforms, along with social media to receive users' comments? (Figure 3).

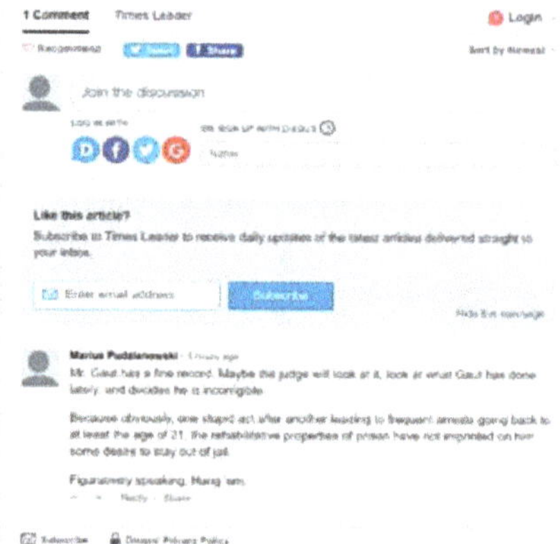

Figure 3. Example acquired from www.timesleader.com.

4. Results

During the walkthrough on the websites, there were a lot of controversies regarding the number of news websites for this study. As a result, there were a lot of changes in order to secure the sample. Throughout the search of the top international news websites, the news aggregators and social media websites were removed from the sample. Additionally, many websites could not be accessed (many of them were from the United States of America) due to the new regulation of the General Data Protection Regulation (GDPR). Consequently, from the 500 most visited news websites of the globe, the final sample was 353. From the overall 353 websites, only 130 allowed users to comment on news articles via their social media accounts (36.82%). In addition, the percentage of news websites that use multiple comment platforms for accepting users' comments was 16.43% (58). Additionally, 64 (18.13%) were using Facebook and 3 (0.84%) were using the commenting platform Disqus (Figure 4) [46].

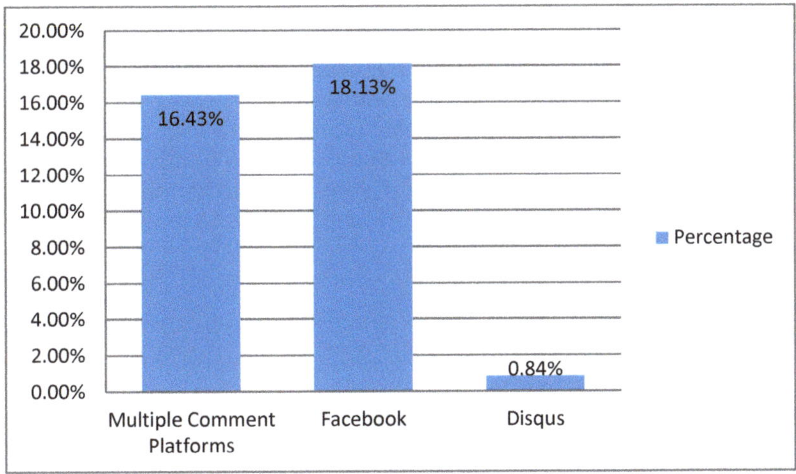

Figure 4. The Results of the Study for the International News Websites for the Commenting Platforms (n = 353).

Regarding the most popular news websites in Greece, the list of 500 top visited news websites of alexa.com was used. Again, the social media platforms and the news aggregators were removed. However, in this case, there were not so many websites excluded due to the GDPR regulations. The main problem with this sample was that the provided list did not include only the news websites of the examined country, but also commercial, marketing, e-shops, and other websites. Consequently, the final number of the examined websites was 175. From the overall 175 websites, only 76 (43.42%) gave to the user, the option of login via their social media accounts and to comment on news articles. On the other hand, the Greek news websites who accepted comments from multiple platforms was 33 (18.85%). The number of news websites that used Facebook was 39 (22.28%), and 2 (1.14%) were those that used the commenting platform Disqus (Figure 5).

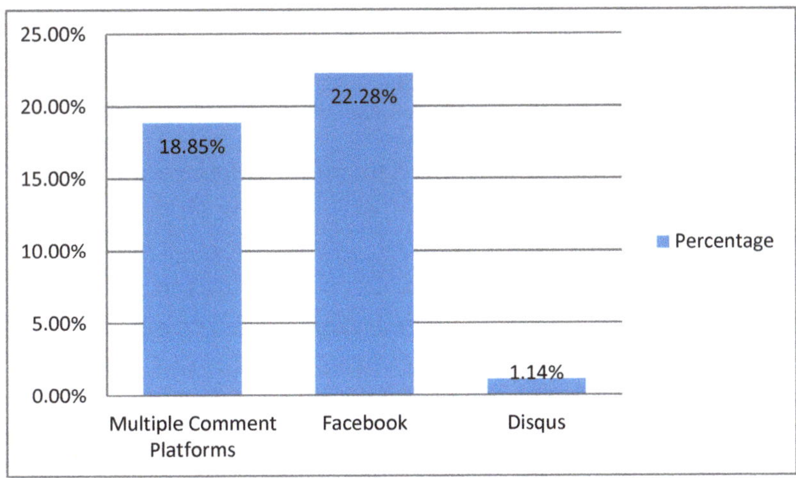

Figure 5. The Results of the Study for the Greek News Websites for the Commenting Platforms (n = 175).

The sample with the news websites of Cyprus had the same problems or characteristics as the previous one. However, there was no problem with the GDPR regulations and no one news website was removed due to this reason. Additionally, an important characteristic emerged from this sample for the Cypriot media system. In the list of the most visited news websites provided by alexa.com, there was a large number of Turkish news websites, which were offering content in both the Turkish and English language. The overall number of the examined news websites was 128. From this number only 51 (39.84%) allowed the users to use their social media accounts in order to comment the provided news articles. The number of news websites with multiple comment platforms was 22 (17.18%). In addition, 26 (20.31%) of the news websites were using Facebook, and only 2 (1.56%) the commenting platforms used Disqus (Figure 6).

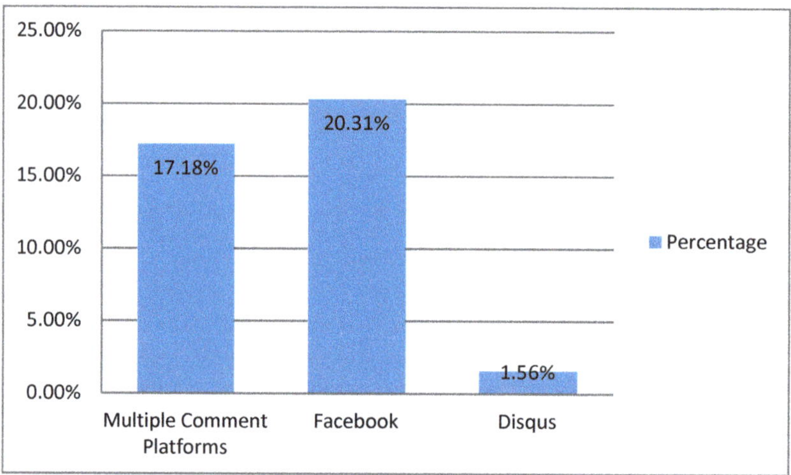

Figure 6. The Results of the Study for the Cypriot News Websites for the Commenting Platforms (n = 128).

5. Discussion

The findings of this study prove that around the world, there seems to be a similar pattern regarding the number of news companies that use multiple comment platforms and social media accounts in order to allow users to comment on their provided content (Figures 7 and 8). Those results revealed, support the arguments of Giomelakis and Veglis [33,40] regarding the practices of news media organizations. The news companies around the world are employing, in some cases, similar techniques to promote their content via commenting. This development seems to be related to the SEO practices that the last decade have become crucial for the financial survival of the news websites. Comments have an important impact in social influence, and information diffusion in relation to users and society. Moreover, the raw and strong content along with the participation of the users via likes, shares, and similar interactions can influence the SEO metrics and improve the popularity of a website [25–27,32].

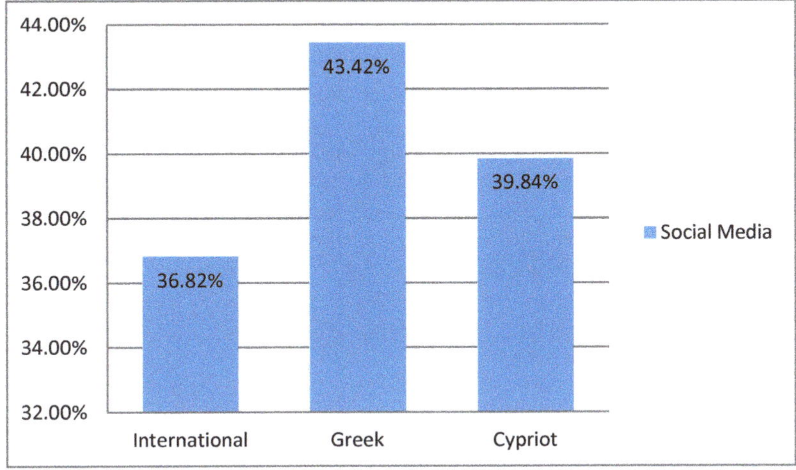

Figure 7. The Results of the Study for the First Research Question.

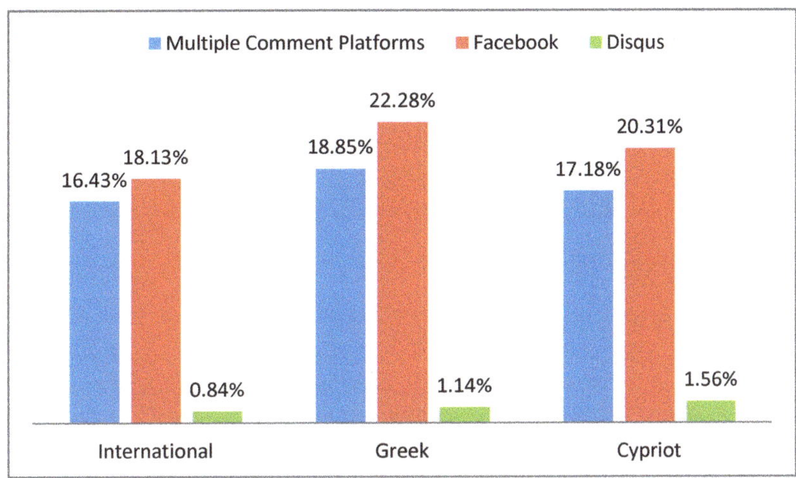

Figure 8. The Results of the Study for the Second Research Question.

Additionally, the current study raises questions about the overall operation of news websites throughout the globe. The majority of the news websites were not using social media accounts or multiple commenting platforms for allowing the users to comment on their news articles (RQ1 and RQ2). These findings are considered significant in the sense that there is no other systematic research until today in the field of journalism and communication that focuses on investigating if the news websites of the globe follow the SEO practices in relation to users' comments.

The aforementioned results for the research questions of this study are probably related to the difficulty of managing the comments of the websites. More specifically, in the few last years there has been a huge debate about the regulation of the comments, as there are cases in which the users promote hate speech, racism, and stereotypes. Social media platforms, such as YouTube and Facebook, were used in the past as a tool from promoting hate against minorities [50]. In this new reality, there is a high chance that the news organizations are afraid of taking the blame for such extreme polarization and for promoting hate. Another important reason for deleting the comments might be the low quality content provided by the users, which is considered to affect, in a negative way, the ranking of a website [28]. The usage of Facebook as a commenting platform is an expected finding because it is supposed to be one of the most popular social media platforms of the world [20].

However, this finding comes against the importance of users' comments as an SEO practice in the sense that the use of a different platform for comments, such as Facebook does not help with improving the relevant Google ranking. On the contrary, it has the opposite effect, as the engagement and interaction are not happening on the website [30]. Moreover, it does not help in building an active community on the website, which would otherwise result again in improving the visibility of the website, as it is considered to be an SEO practice [31]. In addition, the commenting platforms, such as Disqus, can affect, in a good or bad way, the ranking of the website. Because the platform takes time to load the comments, there is a high chance that Google will not index the comments of the website. If the quality is strong, then the website loses visibility, as the high engagement of the users is not evaluated despite it being an SEO practice [29].

Another noteworthy finding of this research is the expansion of the relevant literature for the Greek and Cypriot online media system. One would have thought that there would be major differences between the news websites of the international news organizations and those of Greek and Cypriot news outlets. However, this study argues that all of them share the same features and characteristics, proving that the news websites are being developed based on the same technological knowledge. Perhaps, this finding is also related to the prominence of specific social media platforms, which cannot be easily ignored by the news organizations if they want to attract more unique users, supporting the key roles of social media, social influence, and information diffusion in the contemporary media world.

6. Conclusions and Limitations

This study provides findings for a topic that has not been researched extensively in the field of communication. The SEO practices of the news websites regarding the comments on the websites are supposed to be an essential factor for the popularity of the websites. By commenting, the users provide free of charge, fresh and robust content, which, in some cases, provokes interactions, further engagement, and the users spend more time on the website. Besides, commenting is a form of social influence and information diffusion. Despite improving the visibility of news websites, the comments can even result in changing the opinions of the users.

However, this research shows also that the news websites do not seem to invest a lot of time in improving their visibility via users' comments. By using social media platforms and slow comment systems, they seem to lose the strong content of the users, which is considered to be an important SEO practice. Via the regulation of the comments, the news websites would have been able to maintain strong discussions on their websites and even build communities. Therefore, they would have been ranked higher in the relevant rankings.

Additionally, this study reveals that there are standard practices concerning commenting throughout the world, Greece, and Cyprus. Until today there was no systematic research in the field of journalism and communication that has evaluated the way that users comment on news websites. It seems that a significant number of news websites do not choose to use social media accounts or multiple comment platforms for allowing the users to comment on their news articles. This finding is probably linked with the crucial problem of been blamed for not regulating the harmful content of the website and the low-quality content created by the users. Further, it seems that they are not willing to take the risk of regulating the content in a more efficient way, such as not using slow comment systems and popular social media, such as Facebook. These concerns have to be addressed for increasing the visibility of the news websites, thus the information diffusion and the social influence of the news. Regarding the limitations of this study, it can be argued that the list provided from alexa.com restricted the examined sample and results. The results seem significant in the sense that this research is one of the first studies focusing on the field of news information and communication, but a different list may offer different results from this study. Furthermore, Greece and Cyprus seem to support the way the most popular news websites function, in relation to the comments of the users. However, other countries, for example, those from Asia, may provide different results. Thus, the researchers of this study will try to replicate the same study in Asian countries to further expand, the existing literature. The relationship between SEO and comments on news media websites is a new research topic, and, as a result, there is a lack of research for this specific topic in the field. We believe that for the future, there is a need for a more thorough investigation, via different research methods such as surveys, content analysis, and examining even the traffic of the news media websites.

Author Contributions: Conceptualization, M.-A.K.; methodology, M.-A.K., E.L., M.K., and N.A.; software, M.-A.K., E.L., M.K., and N.A.; validation, M.-A.K.; formal analysis, M.-A.K., E.L., M.K., and N.A.; investigation, M.-A.K., E.L., M.K., and N.A.; resources, M.-A.K., E.L., M.K., and N.A.; data curation, M.-A.K., E.L., M.K., and N.A.; writing—original draft preparation, M.-A.K., E.L., M.K., and N.A.; writing—review and editing, M.-A.K., and N.A.; visualization, M.-A.K., E.L., M.K., and N.A.; supervision, N.A.; project administration, N.A.; funding acquisition, M.-A.K.

Funding: This article is based on research undertaken by the lead author with a doctoral student at Hong Kong Baptist University, supported by the Hong Kong PhD Fellowship Scheme (HKPFS). Apart from that scholarship, the authors received no other financial support for the research, authorship, and/or publication of this article.

Conflicts of Interest: The authors declare no conflict of interest.

References

1. Venkatesh, V.; Morris, M.G.; Davis, G.B.; Davis, F.D. User Acceptance of Information Technology: Toward a Unified View. *MIS Q.* **2003**, *27*, 425. [CrossRef]
2. Asch, S.E. Effects of group pressure upon the modification and distortion of judgments. In *Groups, Leadership, and Men*; Guetzkow, H., Ed.; Carnegie Press: Pittsburgh, PA, USA, 1951; pp. 177–190.
3. Kelman, H.; Hovland, C. Reinstatement of the communication in delayed measurement of attitude change. *J. Abnorm. Soc. Psychol.* **1953**, *48*, 327–335. [CrossRef] [PubMed]
4. Cialdini, R.; Goldstein, N. Social influence: Compliance and conformity. *Annu. Rev. Psychol.* **2003**, *55*, 591–621. [CrossRef] [PubMed]
5. Milgram, S. Behavioural study of obedience. *J. Abnorm. Soc. Psychol.* **1963**, *67*, 371–378. [CrossRef] [PubMed]
6. Latané, B. The psychology of social impact. *Am. Psychol.* **1981**, *36*, 343–356. [CrossRef]
7. Bakshy, E.; Karrer, B.; Adamic, L.A. Social influence and the diffusion of user-created content. In Proceedings of the Tenth ACM Conference on Electronic Commerce—EC'09, Stanford, CA, USA, 6–10 July 2009; ACM Press: New York, NY, USA, 2009; p. 325.
8. Bastiaensens, S.; Pabian, S.; Vandebosch, H.; Poels, K.; Van Cleemput, K.; DeSmet, A.; De Bourdeaudhuij, I. From Normative Influence to Social Pressure: How Relevant Others Affect Whether Bystanders Join in Cyberbullying: Normative Influence on Cyberbullying Bystanders. *Soc. Dev.* **2016**, *25*, 193–211. [CrossRef]
9. Kelman, H. Compliance, identification, and internalization: Three processes of attitude change. *J. Confl. Resolut.* **1958**, *2*, 51–60. [CrossRef]
10. Jalali, M.S.; Ashouri, A.; Herrera-Restrepo, O.; Zhang, H. Information diffusion through social networks: The case of an online petition. *Expert Syst. Appl.* **2016**, *44*, 187–197. [CrossRef]
11. Myers, S.A.; Zhu, C.; Leskovec, J. Information Diffusion and External Influence in Networks. *arXiv* **2012**, arXiv:Physics/1206.1331.
12. Huffaker, D. Dimensions of Leadership and Social Influence in Online Communities. *Hum. Commun. Res.* **2010**, *36*, 593–617. [CrossRef]
13. Susarla, A.; Oh, J.-H.; Tan, Y. Social Networks and the Diffusion of User-Generated Content: Evidence from YouTube. *Inf. Syst. Res.* **2012**, *23*, 23–41. [CrossRef]
14. Peng, S.; Yang, A.; Cao, L.; Yu, S.; Xie, D. Social influence modeling using information theory in mobile social networks. *Inf. Sci.* **2017**, *379*, 146–159. [CrossRef]
15. Xu, W.W.; Park, J.Y.; Kim, J.Y.; Park, H.W. Networked Cultural Diffusion and Creation on YouTube: An Analysis of YouTube Memes. *J. Broadcast. Electron. Media* **2016**, *60*, 104–122. [CrossRef]
16. Otmazgin, N.K. Contesting soft power: Japanese popular culture in East and Southeast Asia. *Int. Relat. Asia Pac.* **2008**, *8*, 73–101. [CrossRef]
17. Nye, J.S. Public Diplomacy and Soft Power. *Ann. Am. Acad. Political Soc. Sci.* **2008**, *616*, 94–109. [CrossRef]

18. Skoric, M.M.; Zhu, Q.; Goh, D.; Pang, N. Social media and citizen engagement: A meta-analytic review. *New Media Soc.* **2015**, *18*, 1817–1839. [CrossRef]

19. Statista. YouTube—Statistics & Facts. 2019. Available online: https://www.statista.com/topics/2019/youtube/ (accessed on 21 August 2019).

20. Statista. Facebook—Statistics & Facts. 2019. Available online: https://www.statista.com/topics/751/facebook/ (accessed on 21 August 2019).

21. Statista. Twitter—Statistics & Facts. 2019. Available online: https://www.statista.com/topics/737/twitter/ (accessed on 21 August 2019).

22. Stieglitz, S.; Dang-Xuan, L. Emotions and Information Diffusion in Social Media—Sentiment of Microblogs and Sharing Behavior. *J. Manag. Inf. Syst.* **2013**, *29*, 217–248. [CrossRef]

23. Schneider, F. China's 'info-web': How Beijing governs online political communication about Japan. *New Media Soc.* **2016**, *18*, 2664–2684. [CrossRef]

24. Xenos, M.A.; Macafee, T.; Pole, A. Understanding variations in user response to social media campaigns: A study of Facebook posts in the 2010 US elections. *New Media Soc.* **2015**, *19*, 826–842. [CrossRef]

25. Carrera, B.; Jung, J.-Y. SentiFlow: An Information Diffusion Process Discovery Based on Topic and Sentiment from Online Social Networks. *Sustainability* **2018**, *10*, 2731. [CrossRef]

26. Winter, S.; Brückner, C.; Krämer, N.C. They Came, They Liked, They Commented: Social Influence on Facebook News Channels. *Cyberpsychologybehaviorand Soc. Netw.* **2015**, *18*, 431–436. [CrossRef] [PubMed]

27. Hong, S.; Cameron, G.T. Will comments change your opinion? The persuasion effects of online comments and heuristic cues in crisis communication. *J. Contingencies Crisis Manag.* **2018**, *26*, 173–182. [CrossRef]

28. Slegg, J. Why Blog Comments Are Great for Google SEO and Users. 2016. Available online: http://www.thesempost.com/comments-good-for-google-seo/ (accessed on 26 August 2019).

29. Yoast. Comment Systems and SEO. 2018. Available online: https://yoast.com/video/ask-yoast-comment-systems-and-seo/ (accessed on 26 August 2019).

30. Schwartz, B. Google: Comments Better On Your Site Then On Social Networks. 2017. Available online: https://www.seroundtable.com/google-comments-better-site-24615.html (accessed on 26 August 2019).

31. Schwartz, B. Google: Community Through Comments Help a Lot with Ranking. 2016. Available online: https://www.seroundtable.com/google-community-help-a-lot-with-ranking-21994.html (accessed on 26 August 2019).

32. Dover, D.; Dafforn, E. *Search Engine Optimization (SEO) Secrets*, 1st ed.; Wiley: Indianapolis, IN, USA, 2011.

33. Giomelakis, D.; Veglis, A. Investigating Search Engine Optimization Factors in Media Websites: The case of Greece. *Digit. Journal.* **2016**, *4*, 379–400. [CrossRef]

34. Barnes, R. The "ecology of participation": A study of audience engagement on alternative journalism websites. *Digit. Journal.* **2014**, *2*, 542–557. [CrossRef]

35. Ksiazek, T.B. Commenting on the News: Explaining the degree and quality of user comments on news websites. *Journal. Stud.* **2018**, *19*, 650–673. [CrossRef]

36. Antonopoulos, N.; Veglis, A.; Gardikiotis, A.; Kotsakis, R.; Kalliris, G. Web Third-person effect in structural aspects of the information on media websites. *J. Comput. Hum. Behav.* **2015**, *44*, 48–58. [CrossRef]

37. Antonopoulos, N.; Veglis, A.; Emmanouloudis, A. Online Marketing for Media: The Case of Greek News Websites. *Int. J. Mark. Stud.* **2017**, *2*, 104–112. [CrossRef]

38. Ziegele, M.; Breiner, T.; Quiring, O. What Creates Interactivity in Online News Discussions? An Exploratory Analysis of Discussion Factors in User Comments on News Items. *J. Commun.* **2014**, *64*, 1111–1138. [CrossRef]

39. Search Engine Land. The Periodic Table of SEO. 2019. Available online: https://searchengineland.com/seotable (accessed on 26 August 2019).

40. Giomelakis, D.; Veglis, A. Employing Search Engine Optimization Techniques in Online News Articles. *Stud. Media Commun.* **2015**, *3*, 22–33. [CrossRef]

41. Hille, S.; Bakker, P. Engaging the Social News User: Comments on news sites and Facebook. *Journal. Pract.* **2014**, *8*, 563–572. [CrossRef]

42. Kalogeropoulos, A.; Negredo, S.; Picone, I.; Nielsen, R.K. Who Shares and Comments on News?: A Cross-National Comparative Analysis of Online and Social Media Participation. *Soc. Media Soc.* **2017**, *3*, 1–12. [CrossRef]

43. Usatoday.com. The Search Engine that Could. 2003. Available online: https://usatoday30.usatoday.com/tech/news/2003-08-25-google_x.htm (accessed on 19 August 2019).

44. Ziakis, C.; Vlachopoulou, M.; Kyrkoudis, T.; Karagkiozidou, M. Important Factors for Improving Google Search Rank. *Future Internet* **2019**, *11*, 32. [CrossRef]

45. Antonopoulos, N.; Veglis, A. Technological Characteristics and Tools for Web Media Companies in Greece. In Proceedings of the 16th Panhellenic Conference on Informatics, Piraeus, Greece, 5–7 October 2012; pp. 44–50, ISBN 978-1-4673-2720-6. [CrossRef]

46. Antonopoulos, N.; Veglis, A. The evolution of the technological characteristics of media websites. In Proceedings of the Asian Conference on Media and Mass Communication, Osaka, Janpan, 8–10 November 2013; pp. 130–146, ISSN 2186-5906. [CrossRef]

47. Allendoerfer, K.; Aluker, S.; Panjwani, G.; Proctor, J.; Sturtz, D.; Vukovic, M.; Chen, C. Adapting the cognitive walkthrough method to assess the usability of a knowledge domain visualization. In Proceedings of the IEEE Symposium on Information Visualization, 2005. INFOVIS 2005, Minneapolis, MN, USA, 23–25 October 2005; pp. 195–202. [CrossRef]

48. Mahatody, T.; Sagar, M.; Kolski, C. State of the Art on the Cognitive Walkthrough Method, Its Variants and Evolutions. *Int. J. Hum. Comput. Interact.* **2010**, *26*, 741–785. [CrossRef]

49. Light, B.; Burgess, J.; Duguay, S. The walkthrough method: An approach to the study of apps. *New Media Soc.* **2018**, *20*, 881–900. [CrossRef]

50. George, C. *Hate Spin: The Manufacture of Religious Offence and its Threat to Democracy*; MIT Press: Cambridge, MA, USA, 2016.

MDPI
St. Alban-Anlage 66
4052 Basel
Switzerland
Tel. +41 61 683 77 34
Fax +41 61 302 89 18
www.mdpi.com

Future Internet Editorial Office
E-mail: futureinternet@mdpi.com
www.mdpi.com/journal/futureinternet

www.ingramcontent.com/pod-product-compliance
Lightning Source LLC
LaVergne TN
LVHW070617170726
843515LV00004B/102